ISNM 72:
International Series of Numerical Mathematics
Internationale Schriftenreihe zur Numerischen Mathematik
Série internationale d'Analyse numérique
Vol. 72

Edited by
Ch. Blanc, Lausanne; R. Glowinski, Paris;
G. Golub, Stanford; P. Henrici, Zürich;
H. O. Kreiss, Pasadena; A. Ostrowski, Montagnola;
J. Todd, Pasadena

Birkhäuser Verlag
Basel · Boston · Stuttgart

Parametric Optimization and Approximation

Conference Held at the Mathematisches Forschungsinstitut, Oberwolfach, October 16–22, 1983

Edited by
B. Brosowski
F. Deutsch

1985

Birkhäuser Verlag
Basel · Boston · Stuttgart

Editors

B. Brosowski
Universität Frankfurt
Fachbereich Mathematik
Robert-Mayer-Str. 10
D–6000 Frankfurt (FRG)

F. Deutsch
Pennsylvania State University
Department of Mathematics
215 McAllister Building
University Park, PA 16802 USA

Library of Congress Cataloging in Publication Data
Main entry under title:

Parametric optimization and approximation.
(International series of numerical mathematics ; vol. 72)
Proceedings of the International Symposium on »Parametric Optimization and Approximation«, »held at the Oberwolfach Research Institute« – – Pref.
1. Mathematical optimization – – Congresses. 2. Approximation theory – – Congresses. I. Brosowski, Bruno. II. Deutsch, F. (Frank), 1936– . III. International Symposium on »Parametric Optimization and Approximation« (1983 : Oberwolfach, Germany) IV. Series: International series on numerical mathematical ; v. 72.
QA402.5.P38 1985 519 85-396
ISBN 3-7643-1671-3

CIP-Kurztitelaufnahme der Deutschen Bibliothek

Parametric optimization and approximation :
conference held at the Math. Forschungsinst., Oberwolfach, October 16–22, 1983 / ed. by B. Brosowski ; F. Deutsch. – Basel ; Boston ; Stuttgart : Birkhäuser, 1985.
(International series of numerical mathematics ; Vol. 72)
ISBN 3-7643-1671-3
NE: Brosowski, Bruno [Hrsg.]; Mathematisches Forschungsinstitut <Oberwolfach>; GT

Printed in Germany
ISBN 3-7643-1671-3

TABLE OF CONTENTS

PREFACE.

This volume contains the proceedings of the International Symposium on "Parametric Optimization and Approximation", held at the Oberwolfach Research Institute, Black Forest, October 16-22, 1983. It includes papers either on a research or of an advanced expository nature. Some of them could not actually be presented during the symposium, and are being included here by invitation. The participants came from Brazil, Bulgaria, CSSR, German Democratic Republic, Great Britain, Israel, Netherlands, South Africa, USA, and West Germany. We take the opportunity to express our thanks to all those who participated in the symposium or contributed to this volume. We also thank the Oberwolfach Mathematical Research Institute for the facilities provided.

August 1984.

Bruno Brosowski, Frankfurt a.M.

Frank Deutsch, University Park, PA

International Series of
Numerical Mathematics, Vol. 72

A RITZ METHOD FOR THE NUMERICAL SOLUTION OF A CLASS OF STATE CONSTRAINED CONTROL APPROXIMATION PROBLEMS

Walter Alt

Mathematisches Institut
Universität Bayreuth
Postfach 3008
D-8580 Bayreuth

Introduction

This paper is concerned with a control approximation problem which occurs in connection with the optimal heating of solids. We consider a one-dimensional homogeneous metal rod which is kept insulated at the left end, and is heated at the right end, where the temperature is regulated by a control function. The problem consists of finding an optimal control such that the deviation of the temperature distribution in the rod at a fixed final time from a desired distribution is minimized; at the same time the temperature has to satisfy certain constraints. We use a Ritz type method to approximate the original problem by a series of discrete convex optimization problems, and we derive error bounds for the extremal values of the discrete problems.

1. The control approximation problem

We consider the following problem:

(P) Minimize $\int_0^1 (y(T,x) - y_T(x))^2\, dx$

subject to $y \in C([0,T] \times [0,1])$, $u \in L^\infty[0,T]$ and

(1.1) $$\frac{\partial y}{\partial t} - \frac{\partial^2 y}{\partial x^2} = 0,$$

(1.2) $$\frac{\partial y}{\partial x}(\cdot,0) = 0,$$

(1.3) $$\alpha y(\cdot,1) + \frac{\partial y}{\partial x}(\cdot,1) = u,$$

(1.4) $$y(0,\cdot) = 0,$$

(1.5) $$\rho_1 \leqslant u(t) \leqslant \rho_2 \qquad \text{a.e. on } [0,T].$$

(1.6) $$y(t,1) \leqslant \eta(t) \qquad \forall\, t \in [0,T].$$

Herein T, α, ρ_1, ρ_2 are given real numbers such that $T > 0$, $\alpha > 0$ and $\rho_1 < \rho_2$; y_T and η are fixed functions with $y_T \in L^2[0,1]$, $\eta \in C[0,T]$.

Let $p \in\,]2,\infty]$ and $u \in L^p[0,T]$ be given. In $L^2[0,1]$ the generalized solution $y(u)$ of (1.1) - (1.4) has the series representation

(1.7) $$y(u)(t) = \sum_{k=1}^{\infty} v_k(1) \int_0^t e^{-\lambda_k(t-\tau)}\, u(\tau)\, d\tau\; v_k \qquad \forall\, t \in [0,T]$$

where the λ_k resp. v_k are the eigenvalues resp. eigenfunctions of the corresponding elliptic eigenvalue problem. The operator S defined by

(1.8) $$S u := (y(u), y(u)(T)) \qquad \forall\, u \in L^p[0,T]$$

is a continuous linear operator from $L^p[0,T]$ into $C([0,T] \times [0,1]) \times L^2[0,1]$ (compare ALT/MACKENROTH [2] and MACKENROTH [5]).

By p_1 resp. p_2 we denote the canonical projection of $C([0,T] \times [0,1]) \times L^2[0,1]$ onto $C([0,T] \times [0,1])$ resp. $L^2[0,1]$. Further we define

(1.9) $K := \{z \in C([0,T] \times [0,1]) \mid z(t,1) \leqslant \eta(t) \ \forall\, t \in [0,T]\}.$

Then problem (P) can be written in the following form:

(PA) Minimize $\|p_2 S u - y_T\|_2^2$
subject to $u \in L^\infty[0,T]$ and

(1.10) $\rho_1 \leqslant u(t) \leqslant \rho_2(t)$ a.e. on $[0,T]$,

(1.11) $p_1 S u \in K.$

2. The Ritz method

For the numerical solution of problem (P) we present a Ritz type method which defines a series of discrete optimization problems approximating the original problem. The original control space is replaced by a finite dimensional one, and the operator S is approximated by a finite series based on (1.7). To this end let for $i \in \mathbb{N}$ numbers $n_i, k_i, m_i \in \mathbb{N}$ be given and decompositions

(2.1)
$$0 = t_0^i < t_1^i < \dots < t_{n_i}^i = T,$$
$$0 = s_0^i < s_1^i < \dots < s_{m_i}^i = T.$$

Let $u_\nu^i : [0,T] \to \mathbb{R}$ be defined by

(2.2)
$$u_\nu^i(t) = \begin{cases} 1 & \text{if } t \in [t_{\nu-1}^i, t_\nu^i] \\ 0 & \text{elsewhere} \end{cases} \qquad (\nu = 1,\dots,n_i)$$

The original control space $U = L^\infty[0,T]$ is replaced by the finite dimensional subspace of piecewise constant functions

(2.3) $U_i := \operatorname{span}\{u_1^i,\dots,u_{n_i}^i\}.$

The operator S is approximated by the finite sum

(2.4)
$$(p_1 S_i u)(t,x) := \sum_{k=1}^{k_i} v_k(1) \int_0^t e^{-\lambda_k (t-\tau)} u(\tau)\, d\tau\; v_k(x).$$

With these notations we can formulate the discrete approximations to (P) resp. (PA) as follows:

(P_i) Minimize $\|p_2 S_i u - y_T\|_2^2$
subject to $u \in U_i$ and

(2.5) $\rho_1 \leqslant u(t) \leqslant \rho_2$ a. e. on $[0,T]$,

(2.6) $p_1 S_i u \in K_i$

where the set K_i is defined by

(2.7) $$K_i = \{z \in C([0,T] \times [0,1]) \mid z(s_\nu^i, 1) \leqslant \eta(s_\nu^i),\ \nu = 0,1,..,m_i\}.$$

Problem (P_i) defines a finite dimensional quadratic optimization problem which can be solved by a suitable numerical procedure.

In order to derive convergence results for the extremal values of the problems (P_i) we need the following Slater condition:

(2.8) There is a control $\bar{u} \in L^\infty[0,T]$ with (1.5) and
$(p_1 S \bar{u})(t,1) < \eta(t) \quad \forall\, t \in [0,T]$.

Let

(2.9)
$$\tau_i := \max\,\{t_\nu^i - t_{\nu-1}^i \mid \nu = 1,\dots,n_i\}$$
$$\sigma_i := \max\,\{s_\nu^i - s_{\nu-1}^i \mid \nu = 1,\dots,m_i\}.$$

In ALT/MACKENROTH [3] we have shown the following result.

Theorem 2.1. Suppose that the Slater condition (2.8) is satisfied and $\lim_{i\to\infty} \tau_i = 0$, $\lim_{i\to\infty} \sigma_i = 0$, $\lim_{i\to\infty} k_i = \infty$. Then there is a number $i_o \in \mathbb{N}$ such that for $i \geqslant i_o$ (P_i) has an optimal solution and

$$\lim_{i\to\infty} \inf\,(P_i) = \inf\,(P).$$

The aim of this paper is to derive in addition error bounds for $|\inf\,(P_i) - \inf\,(P)|$. To this end we use methods similiar to those developed in ALT [1].

3. Error bounds for the extremal values

We start by presenting three auxiliary results which we need for our convergence analysis.

Lemma 3.1. Let U, Z be Banach spaces, $K \subset Z$, $A \in \mathcal{L}(U,Z)$. Suppose that for every $i \in \mathbb{N}$ a subset $K_i \subset Z$ and an operator $A_i \in \mathcal{L}(U,Z)$ are given such that the following conditions are satisfied.

(a) There is an $\bar{u} \in U$ with $A\bar{u} \in \operatorname{int} K$.

(b) $\lim_{i\to\infty} A_i u = Au \quad \forall\, u \in U$.

(c) $K \subset K_i$.

Then there is a real number $\eta > 0$, and for any sequence $\{u_i\} \subset U$ with $\lim_{i\to\infty} u_i = \bar{u}$ there is an $i_o \in \mathbb{N}$ with

$$(3.1) \qquad \eta B_Z \in A_i u_i - K_i \qquad \forall\, i \geqslant i_o.$$

Proof. By (a) there is a $\mu > 0$ with $A\bar{u} - \mu B_Z \subset K$. Let $\{u_i\} \subset U$ be a sequence with $\lim_{i\to\infty} u_i = \bar{u}$. From (b) and the theorem of Banach-Steinhaus we obtain $\lim_{i\to\infty} A_i u_i = A\bar{u}$. Hence, there is an $i_o > 0$ with

$$\|A_i u_i - A\bar{u}\| \leqslant \frac{\mu}{2} \qquad \forall\, i \geqslant i_o.$$

This implies

$$A_i u_i - \eta B_Z \subset K \subset K_i$$

for $\eta := \frac{\mu}{2}$. □

The proof of the following lemma is based on the proof of theorem 2 in ROBINSON [6].

Lemma 3.2. Let U, Z be Banach spaces, $C \subset U$, $K \subset Z$ closed convex sets, and $A \in \mathcal{L}(U,Z)$. Suppose that for every $i \in \mathbb{N}$ closed convex sets $C_i \subset U$, $K_i \subset Z$ and an operator $A_i \in \mathcal{L}(U,Z)$

are given such that assumptions (b), (c) of lemma 3.1 and the following conditions are satisfied:

(a') There is an $\bar{u} \in C$ with $A\bar{u} \in \operatorname{int} K$.

(d) For every $u \in C$ there is a sequence $\{u_i\} \subset U$ with $u_i \in C_i$ and $\lim_{i\to\infty} u_i = u$.

(e) $\|u\| \leq r \quad \forall\, u \in C$ and $\forall\, u \in C_i$ for some constant r.

Define $D \;.= \{u \in U \mid Au \in K\}$ and

$D_i \;.= \{u \in U \mid A_i u \in K_i\}$.

Then there is real number $\eta > 0$ such that the following holds: If $u_o \in C \cap D$ and $\{u_i\} \subset U$ is a sequence with $u_i \in C_i \;\forall\, i \in \mathbb{N}$ and $\lim_{i\to\infty} u_i = u_o$ then there is a $i_o \in \mathbb{N}$ and a sequence $\{v_i\}$ such that for all $i \geq i_o$

(3.2) $\quad v_i \in C_i \cap D_i$

(3.3) $\quad \|v_i - u_o\| \leq \|u_i - u_o\| + \frac{2r}{\eta}\|A_i u_i - A u_o\|.$

<u>Proof</u>. Assumption (d) implies the existence of a sequence $\{\bar{u}_i\}$ with $\bar{u}_i \in C_i$ and $\lim_{i\to\infty} \bar{u}_i = \bar{u}$. Hence by lemma 3.1 there are $\eta > 0$ and $i_o \in \mathbb{N}$ with

$$\eta B_Z \subset A_i\bar{u}_i - K_i \qquad \forall\, i \geq i_o.$$

In particular we have $A_i\bar{u}_i \in \operatorname{int} K \;\forall\, i \geq i_o$.

Now let $u_o \in C \cap D$, a sequence $\{u_i\}$ with $u_i \in C_i$, $\lim_{i\to\infty} u_i = u_o$ and $i \geq i_o$ be given. If $A_i u_i \in K_i$ we choose $v_i \;.= u_i$; this implies $v_i \in C_i \cap D_i$ and $\|v_i - u_o\| = \|u_i - u_o\|$. Hence (3.2) and (3.3) are satisfied. If $A_i u_i \notin K_i$ we set $d_i \;.= d[A_i u_i, K_i]$. Then for any $\delta > 0$ there is a $k_\delta \in K_i$ such that $z_\delta \;.= A_i u_i - k_\delta$ satisfies the inequalities

$$0 < \|z_\delta\| < d_i + \delta.$$

For $\varepsilon \in \,]0,\eta[$ we define

$$z_\varepsilon \;.= -(\eta-\varepsilon)\|z_\delta\|^{-1} z_\delta.$$

It follows $\|z_\varepsilon\| \leq \eta - \varepsilon < \eta$ and therefore $z_\varepsilon \in \eta B_Z$. Hence there exists $k_\varepsilon \in K_i$ with $z_\varepsilon = A_i\bar{u}_i - k_\varepsilon$. With $\lambda := [1 + (\eta-\varepsilon)\|z_\delta\|^{-1}]^{-1}$ we obtain $0 < \lambda < 1$ and

$$(1-\lambda)z_\delta + \lambda z_\varepsilon = \Theta_Z =$$

$$(1-\lambda)(A_iu_i - k_\delta) + \lambda(A_i\bar{u}_i - k_\varepsilon) =$$

$$A_i((1-\lambda)u_i + \lambda\bar{u}_i) - ((1-\lambda)k_\delta + \lambda k_\varepsilon)$$

For $v_i := (1-\lambda)u_i + \lambda\bar{u}_i$, $k_i := (1-\lambda)k_\delta + \lambda k_\delta$ this implies $v_i \in C_i$, $A_iv_i = k_i \in K_i$ and therefore $v_i \in C_i \cap D_i$. Further we have

$$\|v_i - u_o\| \leqslant \|u_i - u_o\| + \|u_i - v_i\|$$

and

$$\|u_i - v_i\| = \|u_i - (1-\lambda)u_i - \lambda\bar{u}_i\| = \lambda\|u_i - \bar{u}_i\|.$$

From $\lambda \leqslant (\eta-\varepsilon)^{-1}\|z_\delta\|$, $\|z_\delta\| < d_i + \delta$ and $\|u_i - \bar{u}_i\| \leqslant 2r$ we get

$$\lambda\|u_i - \bar{u}_i\| \leqslant 2r(\eta-\varepsilon)^{-1}(d_i + \delta).$$

From this we finally obtain

$$\|v_i - u_o\| \leqslant \|u_i - u_o\| + \frac{2r}{\eta-\varepsilon}(d[A_iu_i,K_i] + \delta).$$

By the fact that $A u_o \in K \subset K_i$ the proof is completed by letting δ and ε approach zero. □

<u>Lemma 3.3</u>. Let U, Z be Banach spaces, $C \subset U$, $K \subset Z$ closed convex sets and $A \in \mathcal{L}(U,Z)$. Define $D := \{u \in U \mid Au \in K\}$. Suppose that $u_o \in C \cap D$ and that there is an $\bar{u} \in C$ with $A\bar{u} \in \operatorname{int} K$. Then there is a real number $\eta > 0$ such that for any $u \in U$ there exists $\tilde{u} \in C \cap D$ with

$$\|u-\tilde{u}\| \leqslant \frac{2}{\eta}(\|\bar{u} - u_o\| + \|u - u_o\|)\, d[Au,K]. \tag{3.5}$$

<u>Proof</u>. Define the multivalued function $F : U \to Z$ by

$$F(u) = \begin{cases} Au - K, & u \in C \\ \emptyset, & u \notin C. \end{cases}$$

Then F is a closed convex function and by the assumptions of the theorem there exists $\eta > 0$ with

$$\eta B_Z \subset F(u_o + \|\bar{u} - u_o\| B_U).$$

The assertion of the theorem is therefore a special case of theorem 2 in ROBINSON [6]. □

In order to state our main result we introduce some notations. The cost functional of problem (P) resp. (PA) is denoted by f, i. e.

$$f(u) := \|p_2 S u - y_T\|_2^2.$$

The cost functional of problem (P_i) is denoted by f_i, i. e.

$$f_i(u) := \|p_2 S_i u - y_T\|_2^2.$$

Further, if $p \in]2,\infty]$ is given and q is defined by $\frac{1}{p} + \frac{1}{q} = 1$ for $p < \infty$ and $q = 1$ for $p = \infty$, we can define a continuous function g by

$$(3.6) \qquad g(s) = \sum_{k=1}^{\infty} \lambda_k^{-\frac{1}{q}} (1 - e^{\lambda_k q s}) \qquad \forall\ s > 0.$$

We can now formulate our main result.

Theorem 3.4. Let u* be an optimal solution of (P), and suppose that the Slater-condition (2.8) is satisfied. Let $p \in]2,\infty]$ and a sequence $\{u_i\}$ be given with $\rho_1 \leq u_i(t) \leq \rho_2$ a. e. on [0,T] and $\lim_{i\to\infty} \|u_i - u^*\|_p = 0$. Then there are constants c_1, c_2, c_3, c_4 and an $i_o \in \mathbb{N}$ such that for $i \geq i_o$ the following holds.

(3.7) (P_i) has an optimal solution u_i^*.

(3.8) $f_i(u_i^*) - f(u^*) \leq c_1 \|S_i - S\| + c_2 \|u_i - u^*\|_p$

(3.9) $f(u^*) - f_i(u_i^*) \leq c_3 \|S_i - S\| + c_4\, g(\sigma_i)$
where σ_i is defined by (2.9).

Proof. Assertion (3.7) was shown in ALT/MACKENROTH [4]. To proof (3.8) and (3.9) we define $U = L^p[0,T]$, $C = \{u \in U \mid u$ satisfies (1.5)$\}$, $C_i = C \cap U_i$ and $r = \max\{|\rho_1|, |\rho_2|\}$. Further we will use the fact that there are constants $\tilde{c}_1, \tilde{c}_2$ with

(3.10) $|f_i(u_1) - f(u_2)| \leqslant \tilde{c}_1 \|S_i - S\| + \tilde{c}_2 \|u_1 - u_2\|_p \qquad \forall\, u_1, u_2 \in C.$

By lemma 3.2 there is an $i_o \in \mathbb{N}$ and a sequence $\{v_i\}$ such that for all $i \geqslant i_o$ $v_i \in C_i \cap D_i$ and

$$\|v_i - u^*\|_p \leqslant \|u_i - u^*\|_p + \frac{2r}{\eta} \|p_1 S_i u_i - p_1 S u^*\|$$
$$\leqslant \bar{c}_1 \|u_i - u^*\|_p + \bar{c}_2 \|S_i - S\|.$$

Together with (3.10) and the fact that u_i^* is an optimal solution of (P_i) and v_i is feasible for (P_i) this implies

$$f_i(u_i^*) - f(u_i) \leqslant f_i(v_i) - f(u^*) \leqslant$$
$$\tilde{c}_1 \|S_i - S\| + \tilde{c}_2 \|v_i - u^*\|_p \leqslant$$
$$c_1 \|S_i - S\| + c_2 \|u_i - u^*\|_p$$

with constants c_1, c_2. This proofs (3.8).

Applying lemma 3.3 with $A = p_1 S$ we find $\tilde{u} \in C \cap D$ with

$$\|u_i^* - \tilde{u}\|_p \leqslant \frac{4r}{\eta} d[A u_i, K].$$

With $A_i := p_1 S_i$ we further obtain

$$d[A u_i^*, K] \leqslant \|A_i u_i^* - A u_i^*\| + d[A_i u_i^*, K].$$

Therefore

(3.11) $\|u_i^* - \tilde{u}\|_p \leqslant \bar{c}_3 \|S_i - S\| + d[A_i u_i^*, K]$

with some constant $\bar{c}_3$. In order to estimate $d[A_i u_i^*, K]$ define $z_i(t) := A_i u_i^*(t)$ and

$$z(s) := z_i(s_j) + (s - s_j)(s_{j+1} - s_j)^{-1} (z_i(s_{j+1}) - z_i(s_j))$$
$$\text{for } s \in [s_j, s_{j+1}],\ j = 0, 1, \ldots, m_j - 1.$$

Since $z(s_j) = z_i(s_j)$, $j = 0, 1, \ldots, m_j$ and $z_i(s_j) \leqslant 0$ by the definition of z_i it follows that $z \in K$. Therefore

$$d[A_i u_i^*, K] \leqslant \|z_i - z\|.$$

The proof of lemma 5 in ALT/MACKENROTH [2] shows that $\|z_i - z\| \leqslant 2\, g(\sigma_i)$. This together with (3.11) implies

$$\|u_i^* - \tilde{u}\|_p \leqslant \bar{c}_3 \|S_i - S\| + 2\, g(\sigma_i).$$

Now, using the same arguments as in the proof of (3.8) we obtain

$$f(u^*) - f_i(u_i^*) \leq f(\tilde{u}) - f_i(u_i^*) \leq$$
$$c_3 \| S_i - S \| + c_4 \, g(\sigma_i)$$

with some constants c_3, c_4. This completes the proof. □

Remark. If we choose for u_i the projection of u^* onto U_i then the sequence $\{u_i\}$ satisfies the assumptions of theorem 3.4 and from (3.8) and (3.9) we obtain the result of theorem 2.1.

References

[1] Alt, W. (1984) On the approximation of infinite optimization problems with an application to optimal control problems. Applied Mathematics and Optimization (to appear).

[2] Alt, W. and Mackenroth, U. (1983) Numerical solution of parabolic optimal control problems with an integral state constraint (submitted).

[3] Alt, W. and Mackenroth, U. (1982) On the numerical solution of state constrained coercive parabolic quadratic optimal control problems. Proceeding of the Oberwolfach meeting "Optimale Kontrolle partieller Differentialgleichungen mit Schwerpunkt auf numerischen Verfahren" (to appear).

[4] Alt, W. and Mackenroth, U. (1983) On the computation of optimal solutions of parabolic control problems with a pointwise state constraint (submitted).

[5] Mackenroth, U. (1982) Convex parabolic boundary value problems with pointwise state constraints. Journal for Mathematical Analysis and Applications 87, 256-277.

[6] Robinson, S.M. (1976) Regularity and stability for convex multivalued functions. Mathematics of Operations Research 1, 130-143.

Dr. Walter Alt, Mathematisches Institut, Universität Bayreuth, Postfach 3008, D-8580 Bayreuth, West Germany

International Series of
Numerical Mathematics, Vol. 72

BEST SIMULTANEOUS APPROXIMATION (CHEBYSHEV CENTERS)

Dan Amir
School of Mathematical Sciences, Tel Aviv University, Israel

1. Introduction

The problem of approximating simultaneously a set of data in a given metric space by a single element of an approximating family arises naturally in many practical problems. A common procedure is to choose the "best" approximant by a least squares principle, which has the advantages of existence, uniqueness, stability and easy computability. However, in many cases the least deviation principle makes more sense. Geometrically, this amounts to covering the given data set by a ball of minimal radius among those centered at points of the approximating family. The theory of best simultaneous approximants in this sense, called also Chebyshev centers, was initiated by A. L. Garkavi about twenty years ago. It has drawn more attention in the last decade, but is still in a developing stage. In this short survey I try to describe the main known results and to point at some of the connections between the theory of Chebyshev centers and other problems of Approximation Theory and of Banach Space Theory.

2. Notation and basic definitions

X denotes a real normed linear space. $B(x_0,r) \equiv \{x \in X; ||x - x_0|| \leq r\}$ is the closed r-ball centered at x_0.
$S(x_0,r) \equiv \{x \in X; ||x - x_0|| \leq r\}$ is the r-sphere centered at x_0. The unit ball, $B(0,1)$, is denoted also by B_X or by B, and the unit sphere, $S(0,1)$, by S_X or by S. The dual ball, B_{X^*}, is denoted also by B^*, and its extreme points by ext B^*. A denotes a bounded subset of X,
$r(y,A) \equiv \sup\{||x - y||; x \in A\}$ is the minimal radius of a ball centered at y and covering A. Y denotes a convex subset of X.

$r_Y(A) \equiv \inf\{r(y,A);\ y \in A\}$ is the (relative) Chebyshev radius of A with respect to Y. $r(A) \equiv r_X(A)$ is the (absolute) Chebyshev radius and $r_A(A)$ - the self Chebyshev radius of A. For $\varepsilon \geqslant 0$, let $Z_Y^{\varepsilon}(A) \equiv \{y \in Y;\ r(y,A) \leqslant \leqslant (1+\varepsilon) r_Y(A)\}$. $Z_Y(A) \equiv Z_Y^{0}(A)$ is the (relative) Chebyshev center set of A in Y. Z(A) is the (absolute) Chebyshev center of A(in X). In the case where A is a singleton {x}, the Cehbyshev center $X_Y(A)$ is the ordinary best approximation to x in Y, i.e. the metric projection $P_Y x$.

3. Characterization of the Chebyshev center points

3.1 A basic characterization of the center points is the following

Theorem [3]: Let Y be a convex subset and A a bounded subset of the normed space X. Define, on the dual ball B* with its w*-topology, $A^*(f) \equiv \limsup_{g \to f} \sup_{x \in A} g(x)$. Then $y_0 \in Y$ is in $Z_Y(A)$ if and only if there is $f_0 \in \overline{\text{conv}}\{f \in \text{ext } B^*;\ A^*(f) - f(y_0) = r(y_0,A)\}$ such that $f_0(y_0) = \max_{y \in Y} f_0(y)$.

Remarks. (a) There is such an f_0 which corresponds to all $y_0 \in Z_Y(A)$. (b) If Y is a linear subspace, then necessarily $f_0 \in Y^{\perp}$ (the annihilator of Y in X*). For the absolute center (Y = X) we must have $f_0 = 0$. (c) If Y is n-dimensional then by the Caratheodory and Krein-Milman theorems, f_0 can be taken as a convex combination of n + 1 such extreme points.
(d) B* can be replaced by a norming absolutely convex w*-compact $F \subset B^*$. If we drop the "norming", we get a characterization of the points of Y minimizing $\sup_{x \in A} \sup_{f \in F} f(x - y)$. (e) It follows that $y_0 \in Z_Y(A)$ iff $y_0 \in Z_Y(A \setminus B(y_0,t))$ for some (or:all) $0 \leqslant t < r(y_0,A)$. (f) For sufficiency, it suffices that for every $y \in Y$ there exists some $f \in B^*$ with $A^*(f) - f(y) \geqslant A^*(f) - f(y_0) = r(y_0,A)$.

3.2 In the case when A is compact we have $A^*(f) = \max_{\in A} f(x)$, so that $y_0 \in Y$ is in $Z_Y(A)$ iff there is $f_0 \in \overline{\text{conv}}\{f \in \text{ext } B^*;\ f(x - y_0) = r(y_0,A)$ for some $x \in A\}$ with $f_0(y_0) = \max_{y \in Y} f_0(y)$. Again, the analogous remarks hold.
In particular $y_0 \in Z_Y(A)$ iff $y_0 \in Z_Y(A \cap S(y_0, r(y_0,A)))$.

Laurent and Tuan studied the case of an n-dimensional Y

represented as the intersection of the n-dimensional plane $x_0 + V$ (V a subspace) and the closed half spaces $\{y;\ g(y) \leq w(g)\}$, w a w*-continuous function on a w*-compact $G \subset B^*$. In this case we get $f_0 \in V^{\perp}$ which is a convex combination of n+1 elements from $\{f \in \text{ext } B^*;\ f(y_0 - x) = r(y_0,A)$ for some $x \in A\} \cup \{g;\ g(y_0) = w(g)\}$.

3.3 In the case when <u>X is a Hilbert space</u> and Y is a linear subspace we get, for a compact A and $y_0 \in Y$, $y_0 \in Z_Y(A)$ iff $y_0 \in \overline{\text{conv}}\ P_Y(A \cap S(y_0,r(y_0,A)))$. The noncompact analogue is: $y_0 \in \overline{\text{conv}}\ P_Y(A \setminus B(y_0,t))$ for all $0 < t_0 \leq t < r(y_0,A)$.

3.4 In the case when <u>X = C(T)</u>, T compact Hausdorff, we have $Z_Y(A) = Z_Y(u,v)$ where $u(t) \equiv \liminf_{s\to t} \inf_{x\in A} x(s)$, $v(t) \equiv \limsup_{s\to t} \sup_{x\in A} x(s)$. (If A is compact, then $w(t) = \min_{x\in A} x(t)$, $v(t) = \max_{x\in A} x(t)$ and both are in C(T). Otherwise, we consider Y, A as subsets of $\ell_\infty(T)$). Hence $y_0 \in Y$ is in $Z_Y(A)$ iff it minimizes $\|\frac{v-u}{2} + |\frac{v+u}{2} - y|\ \|$.

If Y is a subspace then by 3.1, $y_0 \in Z_Y(A)$ if $Y^{\perp}$ intersects the w*-closed convex hull of $\{e_t;\ v(t) - y_0(t) = r(y_0,A)\} \cup \{-e_t;\ y_0(t) - w(t) = r(y_0, A)\}$ (where $e_t(x) = x(t)$). In the case dim Y = n this happens iff there is a "straddle point" t_0, i.e. such that $v(t_0)-y_0(t_0) = y_0(t_0) - u(t_0) = r(y_0,A) = r_X(A)$, or there are distinct $t_1,\dots,t_k$, $s_1,\dots,s_m$, $k + m \leq n + 1$, with $v(t_i) - y_0(t_i) = r(y_0,A) = y_0(s_j)$ for $i = 1,\dots,k$, $j = 1,\dots,m$, and $\alpha_i, \beta_j > 0$, so that $\sum_{i=1}^{k} \alpha_i y(t_i) = \sum_{j=1}^{m} \beta_j y(s_j)$ for all $y \in Y$ (If X = C[a,b] and Y is a Haar sybspace, then this implies that k + m = n + 1 and the t_i, s_j interlace, i.e. we have n alternances).

References: [3], [9], [13], [23], [24], [31], [36]. Note that Theorem 2 of [24] is incorrect as stated, as pointed out in [3].

<u>4. Uniqueness of center points</u>

4.1 Uniqueness of centers in Y for all comparts in X is characterized in the following

Theorem. TFAE for a convex subset Y of the normed space X:

(i) $|Z_Y(A)| \leqslant 1$ for every compact $A \subset X$.

(ii) $|Z_Y(A)| \leqslant 1$ for every $A \subset X$ such that every $y \in Y$ has a farthest point in A.

(iii) $|Z_Y(u,v)| \leqslant 1$ for every $u, v \in X$.

(iv) X is strictly convex in every direction of Y, i.e. S_X contains no segment parallel to a segment in Y.

(v) $||u|| = ||v|| = \frac{1}{2}||u + v||,\ u - v \in Y - Y \Rightarrow u = v.$

(vi) Every segment in Y is a Chebyshev set.

In the case Y = X, these just mean that X is strictly convex. The classical non strictly convex spaces have very few subspaces Y enjoying this property, e.g., in $C_o(T)$, $L^1(\mu)$ and $(C[a,b], ||\cdot||_1)$ such subspaces are 1-dimensional (none in $L^1(\mu)$, μ atomless).

4.2 Uniqueness of centers in Y for all bounded subsets of X is characterized by:

Theorem: TFAE for a convex $Y \subset X$:

(i) $|Z_Y(A)| \leqslant 1$ for every bounded $A \subset X$.

(ii) X is uniformly convex in every direction of Y, i.e. for every $z \in Y - Y$, $z \neq 0, \varepsilon > 0$ there is $\delta = \delta(z, \varepsilon) > 0$ such that $||u|| = ||v|| = 1$, $u - v = \lambda z$, $||\frac{u+v}{2}|| > 1 - \delta \Rightarrow |\lambda| < \varepsilon$.

(iii) $||U_n||, ||V_n|| \to 1$, $U_n - V_n = \lambda_n z \neq 0$, $z \in Y\text{-}Y$, $||U_n + V_n|| \to 2 \Rightarrow \lambda_n \to 0$. In the case $Y = X$, these just mean that X is u.c.e.d (uniformly convex in every direction).

4.3 A weaker uniqueness property had been observed first for Haar subspaces of C[a,b] and was extended later to interpolating subspaces. An n-dimensional subspace Y of the normed space X is called interpolating if no nontrivial linear combination of n linearly independent points from extB* annihilates Y, and strictly interpolating if extB* is replaced by its w*-closure.

Theorem [2]: If Y is an interpolating subspace and A is a compact subset of X, and if $r_Y(A) > r(A)$, then $Z_Y(A)$ is a singleton. The analogous result holds for strictly interpolating Y and A bounded.

The same holds if Y is replaced by a "Rozema-Smith" set $\{\sum_{i=1}^{n} c_i y_i ; c_i \in J_i\}$, where $y_1,\ldots,y_n$ are linearly independent elements of X, $J_1,\ldots,J_n$ are intervals of the types: (I) a singleton, (II) a nontrivial proper closed (bounded or unbounded) interval in R, or (III) the whole line R, and such that any subset of $\{y_1,\ldots,y_n\}$ consisting of all y_i with J_i of type III and some y_i's with J_i of type II spans an interpolating (or, respectively, strictly interpolating) subspace.

Rozema and Smith showed that the Haar subspaces and the resulting R-S type sets have the same uniqueness property also in the space $C^1[a,b]$, although it has no interpolating subspaces.

4.4 Uniqueness in C(T). Smith and Ward [38] characterized the subsets having unique absolute center. In the notation of 3.4, this happens iff u,v are continuous and v-u is constant.

For the Haar subspaces Y of C[a,b], the characterization of the situations when $Z_Y(f,g)$ is a singleton is given in [13] by patterns of sign changes.

References: [2],[11],[13],[25],[30],[36],[38]. Note that Theorems 3 and 4 of [30] are incorrect as stated, as pointed out in [2].

5. Existence of centers

5.2 The two basic principles for existence of Chebyshev centers are, essentially, due to Garkavi [25].

The compactness argument. If $Y \subset X$ carries another topology τ such that the norm balls are τ-closed (i.e. such that $y_\alpha \overset{\tau}{\to} y \Rightarrow d(x,y) \leq \liminf d(x,y_\alpha)$), then a τ-accumulation point of a "minimizing sequence" in Y for the Chebyshev radius is necessarily a Chebyshev center. Hence, if Y is boundedly τ-compact, then $Z_Y(A) \neq \emptyset$ for every bounded $A \subset X$. In particular, closed finite-dimensional $Y \subset X$, w-closed Y in reflexive subspaces of X, or w*-closed Y in dual subspaces of X, admit centers for bounded $A \subset X$.

The contraction argument. If P is a (not necessarily linear) contractive projection of the normed space X and $PY \subset Y$, then $PZ_Y(A) \subset Z_{PY}(A)$ for every bounded $A \subset PX$.

Combining this with the compactness argument we get: If the Banach space X is the range of a norm-1 linear projection from its bidual (or, from any dual space containing X), then $Z_X(A) \neq \emptyset$ for every bounded $A \subset X$. In particular, if μ is a σ-finite measure, then $X = L^1(\mu)$ admits centers for bounded subsets. Also, if P is a positive linear contraction in $L^1(\mu)$, then $Y = \{x;Px = x\}$ is the range of a norm-1 projection in $L^1(\mu)$ and admits centers for bounded sets (PROLLA).

5.2 C(T) spaces and the successive approximation argument. While the existence of absolute centers in the classical spaces $L^p(\mu)$ $(1 \leq p \leq \infty)$ is guaranteed by the compactness and contraction arguments, the C(T) spaces case needs a new principle. This was done by Kadets and Zamyatin and others using the Hahn-Tong interposition Theorem or Michael's selection theorem.

The more recent approach of successive approximation has the advantages of simplicity, generality and stability. Call X quasi uniformly convex with respect to its convex subset Y if for every $\varepsilon > 0$ there is $\delta > 0$ such that for every $y,z_0 \in Y$ there is $z \in Y$ with $||z - z_0|| \leq \varepsilon$ and $B(z_0,1) \cap B(y,1-\delta) \subset B(z,1-\delta)$. This enables the successive construction of a Cauchy sequence (z_n) in Y so that $r(z_n,A) \to r_Y(A)$ so that, if Y is closed, $Z_Y(A) \neq \emptyset$ for every bounded $A \subset X$. X is both quasi uniformly convex with respect to Y and strictly convex with respect to Y iff it is uniformly convex with respect to Y, i.e. for every $\varepsilon > 0$ there is $\delta > 0$ such that

$||u_n|| = ||v_n|| = 1$, $u_n - v_n = \lambda_n w \neq 0$, $w \in Y - Y$, $||u_n + v_n|| \to 2 \Rightarrow \lambda_n \to 0$ (This is equivalent to z above being chosen in the segment $[z_0, y]$). However, there are important non strictly convex examples. In particular, if Y is the linear sublattice of C(T) determined by the full set of relations $x(s_\alpha) = \lambda_\alpha x(t_\alpha), s_\alpha, t_\alpha \in T, \lambda_\alpha \geq 0$, and if $\inf \{\lambda_\alpha ; \lambda_\alpha > 0\} = \delta_0 > 0$, then the space $\ell_\infty(T)$ of bounded real-valued functions on T, is quasi uniformly convex with respect to Y, with $\delta(\varepsilon) \geq \delta_0 \varepsilon$. This encompasses the cases Y = C(T), or a closed subalgebra of C(T), $Y = C_0(T)$ (T locally compact) or $C_\sigma(T)$ etc. Also, combining this with the contraction argument, we see that the G-space $X \subset C(T)$ determined by the full set of relations $x(s_\alpha) = \mu_\alpha x(t_\alpha)$, $s_\alpha, t_\alpha \in \Omega$, with $\inf \{|\mu_\alpha| ; \mu_\alpha \neq 0\} > 0$, admits centers for bounded subsets of X.

The argument can be extended to vector-valued functions: If X is uniformly convex or a $C_0(\Omega)$ space (more generally, if X is quasi uniformly convex so that $z = z(y, z_0, \varepsilon)$ can be selected to depend continuously on y), then $\ell_\infty(T,X)$ is quasi uniformly convex with respect to C(T,X) or with respect to any "Stone-Weierstrass" subspace $Y = C(K,X) \cdot \varphi$ where φ is a quotient map of the compact K onto T.

5.3 The M-ideal argument. A closed subspace Y of X is called an M-ideal if there is a projection P of X* onto $Y^\perp$ so that $||f|| = ||Pf|| + ||f - Pf||$ for all $f \in X^*$. This is equivalent to a 3-ball intersection property: If $B(x_i, r_i) \cap Y \neq \emptyset$ for i = 1,2,3 and $\bigcap_{i=1}^{3} B(x_i, r_i) \neq \emptyset$, then also $\bigcap_{i=1}^{3} B(x_i, r_i) \cap Y \neq \emptyset$. In a Lindenstrauss space X (i.e. such that $X^* = L^1(\mu)$ for some μ), this implies that $Z_Y(A) \neq \emptyset$ for every compact $A \subset X$.

5.4 Finite codimensional subspaces of C(T). Garkavi and Zamyatin obtained a complete characterization of the finite codimensional subspaces which admit centers:

Theorem. A. TFAE for a finite codimensional subspace Y of X = C(T).

(i) Y is proximinal in X, i.e. $P_Y x \neq \emptyset \ \forall x \in X$.

(ii) $\forall \varphi \in Y^\perp \ \exists x \in X$ with $||x|| = ||\varphi||$ and $\mu(x) = \varphi(\mu) \ \forall \mu \in Y^\perp$.

(iii) Every $\mu \in Y^\perp$ has a Hahn decomposition of closed sets; $\forall \mu, \mu' \in Y^\perp$, $\mathrm{spt}\mu \setminus \mathrm{spt}\mu'$ is closed and $\mu << \mu'$ on $\mathrm{spt}\mu \cap \mathrm{spt}\mu'$.

(iv) $Z_Y(A) \neq \emptyset$ for all compact $A \subset C(T)$.

B. $Z_Y(A) \neq \emptyset$ for all bounded $A \subset C(T)$ iff it is proximinal and, for every $\mu \in Y^{\perp}$, sptμ is extremally disconnected with respect to T (i.e. the closure of any open $G \subset T$ is relatively open in sptμ). In the case of compact metric T, this means that μ is finitely supported.

5.5 Relative centers and absolute centers in C(T). Smith and Ward [39] showed that, for and $A, Y \subset C(T)$, $r_Y(A) = r(A) + d(Y, Z(A))$. Thus $Z_Y(A) \neq \emptyset$ iff the minimal distance between Y and Z(A) is attained. It should be mentioned here that the absolute center Z(A) is always proximinal in C(T) [23].

5.6 Examples of nonexistence of absolute centers. The first such examples were nonproximinal hyperplanes in C[0,1] which fail to have centers even for triangles, and proximinal hyperplanes in C[0,1] with a continuous annihilator, which fail to have centers for bounded subsets. Even linear sublattices (M-subspaces) of C[0,1] may fail to admit centers for bounded subsets if the inf in 5.2 is 0. This yields also examples of A(S) spaces (affine continuous functions on a compact simplex) which fail to admit centers for bounded subsets.

Instability of the Chebyshev centers in infinite L^1-spaces and other spaces is used to show that c(X) (the space of convergent X-valued sequences) fails to have centers for bounded subsets for $X = \ell^1$ or for a certain 3-dimensional X.

Standard spaces may be renormed to fail to have centers. C[0,1] with the strictly convex equivalent norm $||x|| = ||x||_\infty + ||x||_2$ is so. Every nonreflexive X has an equivalent norm in which some triplet has no center. Most of these counterexamples may be found in [10].

5.7 Proximinal subspaces which admit no centers for pairs. If X is strictly convex but not uniformly convex, then $Z_{c_0}(X)(u,v) = \emptyset$ for some $u,v \in \ell_\infty(X)$. This observation of Yost is used in [8] to show that every infinite dimensional X can be equivalently renormed so that $Z_{c_0}(X)(u,v) = \emptyset$ for some $u,v \in \ell_\infty(X)$.

Garkavi showed that for every nonreflexive closed maximal subspace Y of X, there is an equivalent norm on X in which Y is proximinal, but fails to have a center for some pair of points in X.

References: [8],[10],[25],[27],[28],[29],[32],[33],[38],[39],[41],[42].

6. Continuous dependence on the data

6.1 Local uniform continuity of the Chebyshev center map

Theorem [1]. TFAE for a convex $Y \subset X$:

(i) X is quasi uniformly convex with respect to Y.

(ii) $A \to Z_Y(A)$ is uniformly continuous (with respect to the Hausdorff metric h on 2^X) on every bounded subset of 2^X.

(iii) $A \to Z_Y(A)$ is uniformly continuous on every family of trapezes in B(X).

X is uniformly convex with respect to Y iff Z_Y is also single valued.

Remark: Observe that even in the Euclidean plane the Chebyshev center map is not uniformly continuous on the family of all trapezes. (Consider $\mathrm{conv}((0, \pm R),(R\cos\theta, \pm R\sin\theta))$ and $(\mathrm{conv}((0, \pm R\sin\theta),(R\cos\theta \pm R\sin\theta)))$. On the other hand, it is even globally Lipschitz continuous in the case when the modulus of quasi uniform convexity, $\delta(\varepsilon)$, is proportional to ε, e.g. for the C(T) spaces we get $h(Z(A), Z(G)) \leqslant 2h(A,G)$.

6.2 Upper semicontinuity of the center map. A sufficient condition for upper Hausdorff semicontinuity of the Chebyshev center map $A \to Z_Y(A)$ on a class $\mathcal{F}$ of bounded subsets of X is Mach's condition (P_1):
$\forall A \in \mathcal{F}, \varepsilon > 0 \ \exists \delta > 0$ such that $y \in Z_Y^{\delta}(A) \Rightarrow d(y, Z_Y(A)) < \varepsilon$. All convex and closed finite-dimensional Y in any X, and all w*-closed convex subsets of $X = \ell^1$, have property (P_1) with respect to the class of all bounded subsets of X. Closed convex subsets in locally uniformly convex X have property (P_1) with respect to the compacts in X. If Y is an M-ideal in a Lindenstrauss space X, then it has Mach's property (P_2) with respect to the compacts, i.e. δ depends only on ε and not on the compact A. In this case we get Hausdorff continuity of the Chebyshev center map.

Upper semi continuity of $A \to Z_Y(A)$ on the class of compacts is guaranteed in case X has property (H) (i.e. if weak and norm sequential convergence are equivalent on S_X) and Y is boundedly weakly compact, or if X is a dual space with property (H*) (w* and norm convergence equivalent on S_X) and Y is boundedly w*-sequentially compact. In particular, if X is also strictly convex with respect to Y, then the single-valued map $A \to X_Y(A)$ is continuous on the compact sets.

6.3 Examples of discontinuity of the center map. In infinite-dimensional L^1-spaces, the (absolute) center map is not lower semi-continuous on the class of pairs and has no continuous selection. There is also an example which is 3-dimensional [9]. In another example there the absolute center map is not upper semicontinuous on the family of pairs (in fact it is even not Hausdorff uper semicontinuous).

Discontinuity of the relative center map is less surprising, as we know that this can happen even for ordinary approximation in spaces which are not sufficiently convex. However, even for the Haar subspace span $\{1,t\}$ in $C[0,1]$, there is no continuous selection for the relative center map for pairs.

6.4 Lipschitz continuity on separated sets. Borwein and Keener studied the Lipschitz constants $\lambda_i(X) = \sup\{||Z_A(A) - Z_G(G)||;\ h(A,G) \leqslant 1, (A,G) \in F_i\}$ where F_1, F_2, F_3 are the following families of pairs of closed and convex bounded sets in a u.c.e.d reflexive X:

$F_1 = \{(A,G);\ B(Z_A(A),\ r_A(A)) \cap B(Z_G(G),\ r_G(G)) = \emptyset\}$,

$F_2 = \{(A,\ G);\ Z_A(A) \notin G \text{ and } Z_G(G) \notin A\}$, $F_3 = \{(A,G);\ A \cap G = \emptyset\}$.

They showed that if $\dim X \geqslant 2$ then $\frac{1}{2}(1 + \sqrt{5}) \leqslant \lambda_1(X) \leqslant 2 \leqslant \lambda_3(X)$, with $\frac{1}{2}(1 + \sqrt{5}) = \lambda_1(X)$ for Hilbert space, and that $\lambda_2(X) = \infty$.
If $\dim X \geqslant 3$, then $\lambda_3(X) = \infty$ too. If we replace the self radii and centers $r_A(A)$, $r_G(G)$, $Z_A(A)$, $Z_G(G)$ by the absolute ones, then the corresponding Lipschitz contants $\hat{\lambda}_i(X)$ still satisfy $\frac{1}{2}(1 + \sqrt{5}) \leqslant \hat{\lambda}_1(X)$, $\hat{\lambda}_2(X) = \infty$ for every X, while $\hat{\lambda}_1(X) = \infty$ for $X = (\sum_{n=0}^{\infty} \oplus \ell_{p_n}^{2n+1})_2$ for a suitable $p_n \nearrow 1$.

References: [1], [8], [16], [17], [34], [37], [40].

7. Location and construction of the center, self and absolute radius.

7.1 The center and the convex hull. In the most obvious examples, i.e. in two-dimensional spaces and in inner product spaces, we have $r_A(A) = r(A)$ for every closed and convex bounded subset A, so that the search for a Chebyshev center point can be conducted inside A. However, this is the exception, as shown by

Theorem [26]. TFAE for a Banach space X: (i) X is a Hilbert space or $\dim X \leq 2$. (ii) For every bounded nonempty $A \subset X$, $Z(A) \cap \overline{\text{conv}}\, A \neq \emptyset$. (iii) Same as (ii), with A a triplet $\{x,y,z\}$. (iv) For every $x,y,z \in S_X$, $Z\{x,y,z\} \neq \emptyset$, and if x,y,z are linearly independent, then $r\{x,y,z\} < 1$. (v) For every $x,y,z \in S_X$, $Z\{x,y,z\} \neq 0$, and $0 \in Z\{x,y,z\}$ iff $0 \in \text{conv}\{x,y,z\}$. (vi) $r_A(A) = r(A)$ for every convex $A \subset X$. (vii) $r_G(G) \leq r_A(A)$ for every convex $G \subset A$. (viii) $r_A(A) \leq 1$ for every 2-dimensional section A of B_X.

7.2 Relative centers and the range of the metric projection. By 3.3 we have, for every linear subspace Y of a Hilbert space X and any bounded $A \subset X$, $Z_Y(A) \subset \overline{\text{conv}} \bigcup_{x \in A} P_Y x$. Even this property characterizes Hilbert spaces. In fact, it suffices that $Z_Y\{0,x\} \cap [0,y] \neq \emptyset$ for all 2-dimensional subspaces y of X, $x \in X$ and $y \in P_Y x$. For a fixed (not necessarily 2-dimensional) subspace Y in a densely smooth X (in particular in smooth or separable spaces), this property is equivalent to Y being "centrally symmetric" in the sense of Golomb. However, in the classical spaces such subspaces are quite rare: The only nontrivial centrally symmetric finite-dimensional subspaces of C(T), T compact metric, are the one-dimensional subspaces spanned by functions of modulus 1. For any infinite compact T, C(T) has no finite codimensional centrally symmetric subspaces. The centrally symmetric subspaces of $L^1(\mu)$ are the restriction subspaces [11].

7.3 Construction of centers. While an algorithm for the construction of an absolute center for a finite set in a Euclidean space preceded the theory [43], very little has been done in this direction for other cases, although general methods from optimization theory may be applied. An algorithm for constructing a relative center for a pair {f,g} (hence, for compacts) in C[a,b] was suggested in [12]. It utilizes combinations of weighted and restricted range approximation. In case of nonuniqueness, it yields the nearest element in the center of $\frac{1}{2}(f + g)$.

7.4 Radius vs. self radius. For every convex A in any normed X we have $r(A) \leq r_A(A) \leq 2r(A)$ while, by 7.1, $r(A) = r_A(A)$ characterizes inner product spaces. It is reasonable, therefore, that a nontrivial inequality $\sup\{r_A(A);\ r(A) = 1\} < 2$ is a nontrivial convexity property. This line of investigation is pursued in [7] where this convexity property, shared by all finite-dimensional spaces and by uniformly nonsquare spaces (hence, in particular, by uniformly convex or uniformly smooth spaces) is related to superreflexivity, B-convexity and other Banach space properties, and is shown to be equivalent to the existence of projections of uniformly small norm on all maximal subspaces of X.

7.5 Radius vs. diameter. Even the absolute radius is, in the general case, larger than half the diameter (as shown by Davis, $r(A)=\frac{1}{2}\operatorname{diam} A$ characterizes P_1 spaces, i.e. C(T) with T extremally disconnected). The constant $\sup\{2r(A);\ \operatorname{diam} A = 1\}$ is Jung's constant of X, and is related to the projection constant of X while embedded as a maximal subspace.

The "self Jung constant" $\sup\{2r_A(A);\ A \text{ convex},\ \operatorname{diam} A = 1\}$, is related to the concept of "normal structure" used in the theory of fixed points for nonexpansive mappings. Normal structure of X means that $r_A(A) < \operatorname{diam} A$, and therefore that $Z_A(A)$ is a proper subset of A, for nontrivial convex A. If we start with a w-compact A_o, the transfinite sequence of self centers $A_\alpha = Z_{K_\alpha}(K_\alpha), K_\alpha = \bigcap_{\beta<\alpha} A_\beta$, must then terminate at a single point, the Brodski-Milman center of A_o.

References: [4], [7], [11], [12], [16], [21], [26], [36], [43].

8 Applications of Chebyshev centers.

8.1 Proximinality and minimal projections. The following was conjectured in [19] (also problem 5.6 in Singer's survey): Every finite-codimensional proximinal subspace of a Banach space admits a projection of minimal norm. Motivated by a relation between the projection constant of a maximal subspace M and the Chebyshev radii of the sections of the unit ball by the translates of M, as observed in [22], we construct an example of a proximinal maximal subspace with no minimal projection. Start with a closed and convex A in a normed X for which $Z_X(A) = \emptyset$ (cf. 5.6). We may assume diam A = 2. Let $X_1 = X \times R$ normed by the unit ball $B_1 = \mathrm{conv}\{\pm(A \times \{1\}), B_X \times \{0\}\}$. $M = X \times \{0\}$ is clearly proximinal in X_1. The linear projections P of X_1 onto X are $P(x,\alpha) = (x - \alpha y, 0)$, where $y \in X$ is any. But a simple computation shows that $||P|| = r(y,A)$, so that a minimal norm is not attained.

8.2 Another characterization of Hilbert space. The Garkavi characterization 7.1 and the Hahn - Banach theorem are used in [6] to show the following generalization of Kakutani's chacaterization: A Banach space X of dimension ≥ 3 is Hilbert iff every closed maximal subspace of X admits projections of norm arbitrarily close to 1.

8.3 Optimal recovery. Given normed X,Y,W and linear operators $v:X \to Y$, $u:X \to W$, one looks for a (not necessary linear) operator $e:W \to Y$ which will minimize the maximal estimation error $\sup\{||vx - ew||;\ x \in k, ||ux - w|| \leq \varepsilon\}$ for given $K \subset X, \varepsilon < 0$. This is clearly equivalent to looking for $ew \in Z(v(K \cap u^{-1}B(w,\varepsilon)))$.

8.4 Approximation of functions of several variables by functions of less variables. The following observations are due to Franchetti and Cheney: Let T be a compact Hausdorff space and $Y \subset C(T)$. Then Y is proximinal in $C(T\times V)$ for every topological space V iff Y admits centers for bounded subsets of C(T), and Y is proximinal in $C(T\times K)$ for every compact K iff it admits centers for compact subsets of C(T).

8.5 Other approximation problems can be interpreted as looking for a Chebyshev center, e.g. (i) approximation of a bounded set-valued $\varphi:T\to 2^X$ by a single-valued function (LAU). (ii) approximation of elements from (C(T) by elements of $V\circ\varphi$, where φ is a continuous mapping of T onto K, and $V \subset C(K)$.

9. Generalizations.

9.1 Asymptotic centers. Let (A_α) be a decreasing net of bounded subsets of X. Denote $r^*(y,(A_\alpha)) = \inf_\alpha r(y,A_\alpha)$, $r^*_Y((A_\alpha)) = \inf_{y\in Y} r^*(y,(A_\alpha))$ is the asymptotic radius of (A_α) in Y and $Z^*_Y((A_\alpha)) = \{y\in Y, r^*(y,(A_\alpha)\} = = r^*_Y((A_\alpha))\}$ its asymptotic center in Y. Edelstein considered the case when $A_m = \{x_n; n \geq m\}$ for a given bounded sequence (x_n) in X. Calder, Coleman and Harris [18] considered the case when $A_\alpha = A \setminus F_\alpha$, where A is a given bounded infinite set in X and (F_α) is the net of its finite subsets. Existence of asymptotic centers is studied in [14] and [5]. Asymptotic centers are applied in the theory of fixed points and for problems concerning proximinality of C(T,X) in $\ell_\infty(T,X)$ and, by duality, of the space of compact operators K(X,C(T)) in the space of all bounded linear operators.

9.2 n-nets. The problem of best approximating a bounded set A by n points (n fixed) is not trivial even in the Euclidean plane. Even the best 2-net for a triangle depends on the vertices in a nonstable way. While the compactness and contraction arguments work for n-nets, too. Also, n-nets for compact sets exist in C(T) (an M-ideal argument). However, it turns out that C(T) spaces have bounded subsets which cannot be optimally covered by a pair of balls [8].

9.3 Best simultaneous approximation in other senses. There are many other optimization criteria, cf. e.g. [35].

9.4 The complex case. Some of the arguments used for the results mentioned in this suvey do not carry over to the complex case, cf. e.g.[15].

10. References

1. Amir, D. (1978) Chebyshev centers and uniform convexity. Pacific Jour. Math. 77, 1-6.

2. Amir, D.(1984) Uniqueness of best simultaneous approximation and strictly interpolating subspaces. Jour. Approx. Theory 40 (to appear).

3. Amir, D.(1984) A note on "Approximation of bounded sets". Jour. Approx. Theory (to appear).

4. Amir, D. (1984) On jung's constant and related constants in normed linear spaces. Pacific Jour. Math. (to appear).

5. Amir, C., Deutsch, F. (1979) Approximation by certain subspaces in the Banach space of continuous vector-valued functions. Jour. Approx. Theory 27, 254-270.

6. Amir, D., Franchetti, C. (1983) A note on characterization of Hilbert space, Boll. Unione Mat.Ital. (6) 2-A, 305-309.

7. Amir, D., Franchetti, C. (1984) The radius ratio and convexity properties in normed linear spaces. Trans. Amer. Math. Soc. (to appear).

8. Amir, D., Mach, J. (1983) Best n-nets in normed spaces. Canadian Math. Soc. Conference Proceedings 3, 1-4.

9. Amir, D., Mach, J. (1984) Chebyshev centers in normed spaces. Jour. Approx. Theory 40 (to appear).

10. Amir, D., Mach, J., Saatkamp, K. (1982) Existence of Chebyshev centers, best n-nets and best compact approximants. Trans. Amer. Math. Soc. 271, 513-524.

11. Amir, D., Ziegler, Z.(1980) Relative Chebyshev centers in normed linear spaces, I. Jour. Approx. Theory 29, 235-252.

12. Amir, D., Ziegler, Z. (1981) Construction of elements of the Chebyshev center. Approximation Theory and Appl., Proc. Workshop Technion, Haifa 1980, Academic Press, 1-11.

13. Amir, D., Ziegler (1983) Relative Chebyshev centers in normed linear spaces, II. Jour Approx. Theory 38, 293-311.

14. Anderson, C.K., McKnight, C.K., Hyams, W.H. (1975) Center points of sequences. Canadian Jour. Math. 27, 418-422.

15. Blatt, H.P. (1973), Nicht-lineare gleichmässige Simultanapproximation. Jour. Approx. Theory 8, 210-248.

16. Borwein, J., Keener, L. (1980) The Hausdorff metric and Chebyshev centers. Jour. Approx. Theory 28, 366-376.

17. Bosznay, A.P. (1978) A remark on simultaneous approximation. Jour. Approx. Theory 23, 296-298.

18. Calder, J.R., Coleman, W.P., Harris, R.L. (1973) Centers of infinite bounded sets in a normed space. Canadian Jour. Math. 25, 986-999.

19. Cheney, E.W., Price, K.H. (1970) Minimal projections. Approximation Theory (A. Talbot, editor), Academic Press, 261-289.

20. Davis, W. (1977) A characterization of P_1-spaces. Jour. Approx. Theory 21, 315-318.

21. Franchetti, C. (1977) Restricted centers and best approximation in C(Q). Ann. Fac. Sci. Univ. Nat. Zaire (Kinshasa) 3, 35-45.

22. Franchetti, C. (1983) Projections onto hyperplanes in Banach spaces. Jour. Approx. Theory 38, 319-333.

23. Franchetti, C., Cheney, E.W. (1981) Simultaneous approximation and restricted Chebyshev centers in function spaces. Approximation Theory

and Appl., Proc. Workshop Technion, Haifa, 1980, Academic Press, 65-88.

24. Freilich, J.H., McLaughlin, H.W. (1982) Approximation of bounded sets. Jour. Approx. Theory 34, 146-156.

25. Garkavi, A.L. (1962) The best possible net and the best possible cross section of a set in a normed space. Izv. Akad. Nauk SSSR 26, 87-106 (Amer. Math. Soc. Transl. 39 (1964), 111-132).

26. Garkavi, A.L. (1964) On the Chebyshev center and the convex hull of a set. Uspehi Mat. Nauk 19, 139-145.

27. Garkavi, A.L. (1973) The conditional Chebyshev center of a compact set of continuous functions. Math. Notes 14, 827-831.

28. Garkavi, A.L., Zamyatin, V.N. (1975) Conditional Chebyshev center of a bounded set of continuous functions. Math. Notes 18, 622-627.

29. Kadets, M.I., Zamyatin, V.N. (1968) Chebyshev centers in the space C[a,b]. Teor. Funk., Funkc. Anal. Priložz. 7, 20-26.

30. Lambert, J.M., Milman, P.D. (1979) Restricted Chebyshev centers of bounded subsets in arbitrary Banach spaces. Jour. Approx. Theory 26, 71-78.

31. Laurent, P.J., Tuan, P.D. (1970) Global approximation of a compact set in a normed linear space. Numer. Math. 15, 137-150.

32. Mach, J. (1979) Best simultaneous approximation of bounded functions with values in certain Banach spaces. Math. Annal. 240, 157-164.

33. Mach, J. (1979) On the existence of best simultaneous approximations. Jour. Approx. Theory 25, 258-265.

34. Mach, J. (1980) Continuity properties of Chebyshev centers. Jour. Approx. Theory 29, 223-230.

35. Milman, P.D. (1977) On best simultaneous approximation in normed linear spaces. Jour. Approx. Theory 20, 223-238.

36. Rozema, E.R., Smith, P.W. (1976) Global approximation with bounded coefficients. Jour. Approx. Theory 16, 162-174.

37. Sastry, K.P.R., Naidu, S.V.R. (1979) Upper semi continuity of the simultaneous approximation operator. Pure Appl. Math. Sci. (India) 10, 7-8.

38. Smith, P.W., Ward, J.D. (1975) Restricted centers in subalgebras of C(X), Jour. Approx. Theory 15, 54-59.

39. Smith, P.W., Ward, J.D. (1975) Restricted centers in C(Ω). Proc. Amer. Math. Soc. 48, 165-172.

40. Szeptycki, P., van Vleck, F.S. (1982) Centers and nearest points of sets. Proc. Amer. Math. Soc. 85, 27-31.

41. Ward, J. (1974) Chebyshev centers in spaces of continuous functions. Pacific Jour. Math. 52, 283-287.

42. Zamyatin, V.N. (1973) Relative Chebyshev centers in the space of continuous functions. Soviet Math. (Dokl.) 14, 610-614.

43. Zuhovicki, S.I. (1951) An algorithm for finding the point of least deviation (in the sense of P. Chebyshev) from a given system of m points. Dopov. Akad. Nauk. Ukrain. RSR, 404-407.

Prof. Dan Amir, School of Mathematical Sciences, Tel Aviv University, Ramat Aviv, 69978 Tel-Aviv, Israel.

International Series of
Numerical Mathematics, Vol. 72

CHARACTERIZATION OF STRONG UNICITY IN SEMI-INFINITE OPTIMIZATION BY CHAIN OF REFERENCES

Hans-Peter Blatt

Mathematisch-Geographische Fakultät, Katholische Universität Eichstätt, D-8078 Eichstätt, Federal Republic of Germany

We use as in generalizations of the Remez algorithm chain of references instead of references introduced by STIEFEL [8] for obtaining a new characterization of strongly unique optimal solutions.

1. The minimization problem

Let T be a compact set and

$$a: T \to \mathbb{R}^n, \quad f: T \to \mathbb{R}$$

be two continuous mappings. Denoting by $\langle\cdot,\cdot\rangle$ the scalar product in $\mathbb{R}^n$, we consider the continuous convex functional

(1.1) $$g(x) := \max_{t \in T} \big(< a(t), x > - f(t) \big).$$

The problem consists in <u>minimizing</u> $g(x)$ <u>with</u> <u>respect</u> <u>to</u> $x \in \mathbb{R}^n$. We suppose that the minimal value

(1.2) $$\alpha := \inf_{x \in \mathbb{R}^n} g(x)$$

is finite.

<u>Example:</u> Let E be a linear space over $\mathbb{C}$ with norm $\|\cdot\|$, $V = \text{span}\,\{v_1, v_2, \ldots, v_n\}$ a n-dimensional subspace of E. For a fixed element $f \in E$ we want to find an element $\overline{v} \in V$ such that

(1.3) $$\min_{v \in V} \|f-v\| = \|f-\overline{v}\|.$$

If S_{E^*} is the unit cell of the dual space E*, it is well-known that

$$\|h\| = \max_{L \in S_{E^*}} \text{Re}\, L(h)$$

for all $h \in E$. Consider for $v \in V$ the representation

$$v = \sum_{k=1}^{n} (x_k + i\, x_{k+n})\, v_k$$

and define

$$x := (x_1, \ldots, x_{2n}),\quad x_k \in \mathbb{R} \text{ for } 1 \leqq k \leqq 2n,$$

$$R(L) := \big(\text{Re}\, L(v_1), \ldots, \text{Re}\, L(v_n)\big),$$

$$I(L) := \big(-\text{Im}\, L(v_1), \ldots, -\text{Im}\, L(v_n)\big),$$

$$a(L) := \big(R(L), I(L)\big),$$

$$f(L) := \text{Re}\, L(f).$$

Then

$$\text{Re } L(v-f) = < a(L),x > - f(L)$$

and the problem (1.3) is equivalent to minimize the functional

$$g(x) := \max_{L \in S_{E^*}} \left(< a(L),x > - f(L)\right)$$

with respect to $x \in \mathbb{R}^{2n}$. We note that $a: S_{E^*} \to \mathbb{R}^{2n}$ and $f: S_{E^*} \to \mathbb{R}$ are continuous mappings and S_{E^*} is compact if E^* is endowed with the weak topology.

In the following let $||\cdot||$ denote a norm in $\mathbb{R}^n$.

Definition: A solution $x^* \in \mathbb{R}^n$ of the minimization problem is called strongly unique, if there exists a constant $\gamma > 0$ such that

$$g(x) \geqq g(x^*) + \gamma ||x-x^*||$$

for all $x \in \mathbb{R}^n$.

Brosowski [4] proved a characterization of strongly unique optimal solutions by using primitive extremal signatures. This paper is motivated to give a characterization by using notions known from generalizations of the Remez algorithm [2], [5],[6].

2. Chain of references: The exchange theorem

Let V be a subspace of $\mathbb{R}^n$.

Definition: A subset $R = \{a(t_i) \mid 1 \leqq i \leqq k+1, t_i \in T\}$ is called a V-reference, if

(α) there exist $\lambda_i > 0$ ($1 \leqq i \leqq k+1$) such that

$$\sum_{i=1}^{k+1} \lambda_i = 1 \quad \text{and} \quad \sum_{i=1}^{k+1} \lambda_i \, a(t_i) \in V^{\perp},$$

where $V^{\perp}$ is the orthogonal complement of V in $\mathbb{R}^n$, (β) R is minimal, i.e. there exist no proper subset of R such that (α) holds.

The numbers λ_i are uniquely determined and called characteristic numbers of the V-reference R.

Definition: Let W be an affine subspace parallel to V, R a V-reference. Then $x \in W$ is called solution of the V-reference R in W, if

$$\min_{z \in W} \max_{1 \leq i \leq k+1} \left(< a(t_i), z > - f(t_i) \right) = \max_{1 \leq i \leq k+1} \left(< a(t_i), x > - f(t_i) \right).$$

For constructing such a solution we define

$$h_R := \sum_{i=1}^{k+1} \lambda_i \left(< a(t_i), z > - f(t_i) \right)$$

for a fixed $z \in W$ and obtain $y \in V$ by solving the linear system

$$< a(t_i), y > = f(t_i) - < a(t_i), z > + h_R$$

for $1 \leq i \leq k+1$. Then $x = z + y$ is a solution of the V-reference R in W, in general not unique. h_R is called deviation of the V-reference R in W.

Consider now the following construction: Let R_1 be a V-reference with $V = \mathbb{R}^n$, W_1 the set of solutions of the $\mathbb{R}^n$-reference R_1 in $W = \mathbb{R}^n$, h_1 the deviation of R_1 in $\mathbb{R}^n$,

$$V_1 := [R_1]^{\perp},$$

where we denote for abbreviation by $[R_1]$ the subspace in $\mathbb{R}^n$ spanned by the vectors of R_1. Then $\dim V_1 = n-k_1$, if R_1 has k_1+1 elements, and W_1 is parallel to V_1.

We repeat the same construction for $V = V_1$ and $W = W_1$: Let R_2 be a V_1-reference, W_2 the set of solutions of the V_1-reference R_2 in W_1, h_2 the deviation of R_2 in W_1,

$$V_2 := [R_2]^{\perp} \cap V_1 .$$

Then $\dim V_2 = n - k_1 - k_2$, if R_2 has $k_2 + 1$ elements, and W_2 is parallel to V_2, etc...

In this way we possibly can find a <u>chain of references</u>

$$\underline{R} = (R_1, R_2, \ldots, R_s)$$

with corresponding subspaces

$$V_1 \supset V_2 \supset \ldots \supset V_s$$

and parallel affine subspaces

$$W_1 \supset W_2 \supset \ldots \supset W_s,$$

and <u>a vector of deviations</u>

$$h = (h_1, h_2, \ldots, h_s)$$

such that

(1.4) R_i is a V_{i-1}-reference,

(1.5) $V_i = [R_i]^{\perp} \cap V_{i-1}$, W_i is the set of solutions R_i in W_{i-1},

(1.6) h_i is the deviation of R_i in W_{i-1},

(1.7) $V_s = [0]$

$(V_0 := W_0 := \mathbb{R}^n)$.

Such a chain of references determines a unique point $x_{\underline{R}}$ in $\mathbb{R}^n$ $(\{x_{\underline{R}}\} = W_s)$, which is the <u>solution of the chain $\underline{R}$ in $\mathbb{R}^n$</u>.

<u>Definition:</u> $\underline{R}$ is called <u>regular</u>, if each R_i has at least two elements.

If $\underline{R}$ is not regular we get a regular chain by cancelling in $\underline{R}$ each R_j having only one element. Thereby the solution of the chain is not changed.

One of the main tools of generalizations of the Remez algorithm is the <u>exchange theorem</u> between two consecutive references [2], [5]: Let

$$\underline{R} = (R_1, \dots, R_j, R_{j+1}, \dots, R_s)$$

be a chain of references with deviation vector

$$h = (h_1, \dots, h_j, h_{j+1}, \dots, h_s).$$

Then it is possible "to exchange R_j and R_{j+1}" such that we get a new chain

$$\underline{\tilde{R}} = (R_1, \dots, R_{j-1}, \tilde{R}_j, \tilde{R}_{j+1}, R_{j+2}, \dots, R_s)$$

with the following properties:

(1.8) $\quad R_{j+1} \subset \tilde{R}_j,$

(1.9) $\quad \tilde{R}_{j+1} \subset R_j$ and $\tilde{R}_{j+1} \neq \emptyset,$

(1.10) $\quad \tilde{h}_j = (1-\delta)\, h_j + \delta \cdot h_{j+1}$ with $0 < \delta \leqq 1.$

3. Characterization of strong unicity

Let us define for $x \in \mathbb{R}^n$ the set $M(x)$ by

$$M(x) := \{a(t) \mid t \in T,\ g(x) = \langle a(t), x \rangle - f(t)\}.$$

Then $M(x) \neq \emptyset$, $M(x)$ is compact, and we have the following characterization of strongly unique optimal solutions.

Theorem: A point x^* is a strongly unique optimal solution, if and only if $M(x^*)$ contains a chain $\underline{R}$ of references.

Proof: Let us assume that $M(x^*)$ contains a chain $\underline{R}$ of references,

$$\underline{R} = (R_1, R_2, \ldots, R_s).$$

We may assume that $\underline{R}$ is regular. Let

$$R_1 = \{a(t_i) \mid 1 \leq i \leq k+1\}$$

and let us denote by $\lambda_1, \ldots, \lambda_{k+1}$ the characteristic numbers associated with $a(t_1), \ldots, a(t_{k+1})$. Then for each $1 \leq i \leq k+1$ we have

$$g(x) \geq \langle a(t_i), x^* \rangle - f(t_i) + \langle a(t_i), x - x^* \rangle$$

and consequently

$$g(x) \geq \alpha + \max_{1 \leq i \leq k+1} \langle a(t_i), x - x^* \rangle. \tag{3.1}$$

We assert that

$$K_1 := \max_{1 \leq i \leq k+1} \langle a(t_i), x - x^* \rangle \geq \frac{\lambda}{1-\lambda} K_2 \tag{3.2}$$

with

$$K_2 := \max_{1 \leqq i \leqq k+1} | < a(t_i), x-x^* > |$$

and $\lambda := \min_{1 \leqq i \leqq k+1} \lambda_i > 0$. We have only to consider the case that $K_1 < K_2$. Then there exists an index i_0 with

$$K_2 = - < a(t_{i_0}), x-x^* >.$$

Since

$$0 = < \sum_{i=1}^{k+1} \lambda_i \; a(t_i), x-x^* >$$

$$= - \lambda_{i_0} \cdot K_2 + \sum_{i \neq i_0} \lambda_i < a(t_i), x-x^* >$$

we have got

$$K_2 = \frac{1}{\lambda_{i_0}} \cdot \sum_{i \neq i_0} \lambda_i < a(t_i), x-x^* >$$

$$\leqq \frac{1-\lambda}{\lambda} K_1.$$

Hence (3.2) is proved and from (3.1):

(3.3) $$g(x) \geqq \alpha + \frac{\lambda}{1-\lambda} \max_{1 \leqq i \leqq k+1} | < a(t_i), x-x^* > |.$$

Using the exchange theorem we exchange R_2 with R_1 and get an analogous estimation as in (3.3), where the vectors $a(t)$ of R_2 are involved on the right-hand side of (3.3). By the same method it is possible to exchange the vectors of any R_j up to the first component of a chain of references. Finally by summarizing all inequalities of type (3.3) we get with a positive constant γ independent of x:

(3.4) $$g(x) \geqq \alpha + \gamma \max_{a(t) \in S} | < a(t), x-x^* > |$$

with $S = \bigcup_{i=1}^{s} R_i$. (3.4) is equivalent to the strong unicity of x^*, since by (1.7) $[S] = \mathbb{R}^n$ and therefore

$$\max_{a(t) \in S} | < a(t), x > |$$

is a norm in $\mathbb{R}^n$.

Conversely let us assume that there exists no chain of references in $M(x^*)$. As described in section 2 we can construct a sequence

$$(R_1, R_2, \ldots, R_s)$$

with $R_i \subset M(x^*)$ for $1 \leqq i \leqq s$, satisfying the conditions (1.4) - (1.6), but not the condition (1.7); namely $V_s \neq [0]$. Moreover it is impossible to find any V_s-reference R_{s+1} in the set

$$A := M(x^*) \setminus \bigcup_{i=1}^{s} R_i$$

with $|R_{s+1}| > 1$. Hence, if we denote by ri conv (A) the relative interior of the convex hull of A, we have

$$\text{(3.5)} \qquad \text{ri conv } (A) \cap V_s^{\perp} = \emptyset.$$

Using separation theorems for convex sets [7] we conclude that there exists a hyperplane separating properly conv (A) and $V_s^{\perp}$, i.e. there exists a vector $b \in \mathbb{R}^n$ such that

$$< z, b > \leqq 0 \text{ for all } z \in A$$

and

$$< y, b > \geqq 0 \text{ for all } y \in V_s^{\perp},$$

whence $b \in V_s$. Summarizing we have got a vector $b \in V_s$ such that

$$\sup_{a(t)\in A} \langle a(t), b \rangle \leqq 0.$$

But since $V_s = [\bigcup_{i=1}^{s} R_i]^{\perp}$ we have

$$\max_{a(t)\in M(x^*)} \langle a(t), b \rangle \leqq 0,$$

and therefore x^* cannot be strongly unique.

Finally we remark that the method outlined in the first part of the proof can be used to get a posteriori error estimations for optimal solutions, if approximations are calculated by generalized algorithms of the Remez type [3].

4. References

1. Bartelt, M.W. and H.W. McLaughlin (1973) Characterizations of strong unicity in approximation theory, J. Approximation Theory 9, 255-266.

2. Blatt, H.-P., U. Kaiser and B. Ruffer-Beedgen (1983) A multiple exchange algorithm in convex programming, in "Optimization: Theory and algorithms" (J.B. Hiriart-Urruty, W. Oettli, J. Stoer (ed.)), Marcel Dekker, New York, 113-130.

3. Blatt, H.-P., Exchange algorithms, error estimations and strong unicity in convex programming and Chebyshev approximation, to appear in Proceedings of the NATO-Advanced Study Institute, St. John's, Newfoundland (Canada), August 1983.

4. Brosowski, B. (1984) A refinement of an optimality criterion and its application to parametric programming, J. of Optimization Theory and Appl. 42, 367-382.

5. Carasso, C. and P.J. Laurent (1978) Un algorithme de minimisation en chaine en optimisation convexe, SIAM J. Control and Optimization 16, 209-235.

6. Carasso, C. and P.J. Laurent (1978), An algorithm of successive minimization in convex programming, R.A.I.R.O., Analyse numérique, Numerical Analysis 12, 377-400.

7. Rockafellar, R.T. (1970) Convex Analysis, Princeton University Press, Princeton.

8. Stiefel E. (1959) Über diskrete und lineare Tschebyscheff-Approximation, Numer. Math. 1, 1-20.

Hans-Peter Blatt, Mathematisch-Geographische Fakultät,
Katholische Universität Eichstätt, Ostenstraße 18,
D-8078 Eichstätt, Federal Republic of Germany.

International Series of
Numerical Mathematics, Vol. 72

APPLICATION OF PARAMETRIC PROGRAMMING TO THE OPTIMAL DESIGN OF STIFFENED PLATES

Bruno Brosowski

Fachbereich Mathematik der Johann Wolfgang Goethe-Universität, Frankfurt am Main, West Germany

1. Introduction

Stiffened plates are used in various areas of engineering. Examples of their use are f.e. box girder bridge , bridge decks and in ship construction. The use of stiffeners will increase the collapsing load of the structure without increasing the weight and the costs to much. A typical example of a plate stiffened in one direction is shown in figure 1. The designer of such structures aims to keep the weight and the costs of the stiffened plate as low as possible and the collapsing load as high as possible. Thus, the designer is confronted with the problem of satisfying a number of conflicting objectives, i.e. with a vector optimization problem of the following type:

VOP. Minimize the objective functions

$$p_1(x),\ p_2(x),\ p_3(x)$$

subject to the side conditions

$$\forall_{j \in J} \quad A_j(x) \leq 0.$$

The constraints $A_j(s) \leq 0$, j in an index set J, represent the constraints on the buckling stresses and any constraints on the geometrical variables. The vector $x := (x_1, x_2, \ldots, x_l)$ denotes the vector of the design variables. Since the stiffened plate forms part of a more extensive structure, we can assume that the plate has a given width and a given length. Thus, the design variables are the thickness of the plate and the number, depth, and the thickness of the stiffeners.

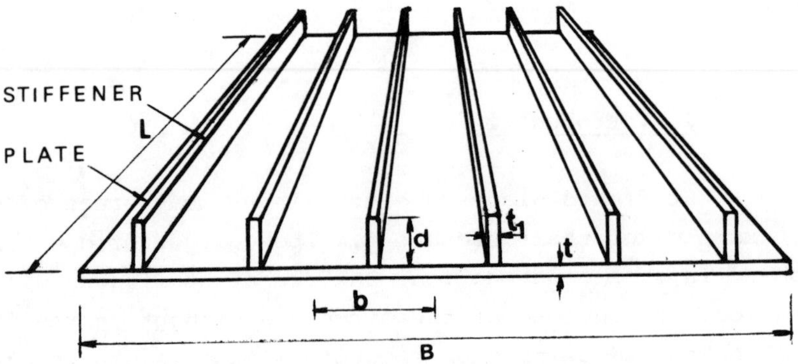

Fig.1. A TYPICAL STIFFENED PLATE

It is well-known that, in general, there does not exist a point x_o in the feasible set

$$Z := \bigcap_{j \in J} \{x \in \mathbb{R}^l \,|\, A_j(x) \leq 0\}$$

such that each of the objective functions p_1,p_2,p_3 attains its infimum in x_o. As usually, we introduce the concept of an *"efficient point"*, compare f.e. GUDDAT [3] or ZELENY [10]:

A point $x_o \in Z$ is called p-efficient, if and only if

$$\forall_{x \in Z} \quad p(x) \leqq p(x_o) \implies p(x) = p(x_o) \quad ,$$

where we have used the abbreviation

$$p(x) := (p_1(x), p_2(x), p_3(x)) .$$

By the condition of efficiency, usually a subset of the feasible set is defined, which may be very large. Also, in contrast to the non-uniqueness in scalar optimization, the vector valued objective function may not be constant on the set of efficient points. If the designer's preference function is unknown or to complex, then it is desirable to determine the whole set of all efficient points or to develop an interactive procedure, which determines successively a sequence of efficient points according to the preferences of the designer.

For the computation of the efficient points, the vector optimization problem can be reduced to a parametric optimization problem using a new scalarization developed in [2]. This led to a new numerical technique, which was tested in the special case of minimization of the weight and of maximization of the collapsing load of a plate stiffened by rectangular stiffeners, compare [1]. Further it was assumed in [1] that the plate has a given thickness.

In this paper we continue the investigations begun in [1,2]. First we state the optimal design. An investigation of the side conditions leads to a simplification of the numerical method developed in [2]. The optimal design problem for weight minimization and for maximization of the collapsing load is considered explicitely in the cases of constant and of variable thickness of the plate. Further we discuss the necessary modifications of the numerical methods, if we also take into account the minimization of the costs.

2. List of symbols

a_o : initial deflection,
A : area of cross section,
A' : area of reduced cross section,
b : spacing of the stiffeners (Figure 1),
b_e : effective width,
B : width of the panel (Figure 1),
d : depth of the stiffener (Figure 1),
E : Young's modulus,
I : moment of inertia of A,
I' : moment of inertia of A',
L : length of the panel (Figure 1)
N : number of stiffeners in a panel,
P_f : axial load at failure,
r : radius of gyration of cross section A,
r' : radius of gyration of cross section A'
t : thickness of the plate,
t_1 : thickness of the stiffener,
y_1 : distance from neutral axis to outer fibre,
y_1' : distance from neutral axis of reduced stiffenend plate to outer fibre,
Δ' : shift of position of neutral axis as plate buckles scaled back to zero axial load,
η' : imperfection parameter of a reduced stiffened plate,
ν : Poisson's ratio,
μ : density of the material,
σ_{cr} : critical stress of a pin sided plate,
σ_{cr2} : critical stress of plate free on one side and pinned on the other three sides,
σ_e : Euler stress of a simple strut,
σ_e' : Euler stress of a reduced stiffened plate,
σ_f' : average axial stress at collapse of reduced stiffened plate,
σ_m' : average axial stress in plate, when it fails,
σ_y : yield stress.

3. Formulation of the optimal design problem

The panel studied is shown in Figure 1. It is designed such that the weight, costs and the negative collapsing load have an efficient point. The design variables are the thickness of the plate and the number N, the depth d and the thickness t_1 of the stiffeners. Using the notation of Figure 1 the weight of the stiffened plate is given by

(P1) $$W := W(t,N,t_1 d) = (bt + dt_1)\mu NL.$$

The cost of fabrication of a stiffened plate is the sum of the cost of the material C_M and that of the welding C_W. Then one has

$$C = C_M + C_W \ ,$$

where $C_M = W \cdot k$ and k denotes the cost of one unit. The real cost for welding is a very complicated function of many variables. It can be approximated by

$$Q(t_{min}) = 0.2133 \cdot 10^{-3} t_{min}^4 + 0.695 \cdot 10^{-3} t_{min}^3 + 0.108 \cdot 10^{-3} t_{min}^2 + 0.226 \cdot t_{min} + 0.33 [DM/m],$$

where $t_{min} = \min(t,t_1)$. Thus, we obtain $C_W = Q(t_{min})NL$ and consequently

(P2) $$C = C_M + C_W = Wk + Q(t_{min})NL = (tB + Nt_1 d)L\mu k + Q(t_{min})NL.$$

(For a discussion compare the papers of LAWO et al.[4] and VESTER [8]).

The collapsing load is calculated approximately by using a proposal of MURRAY [6]. If we make the assumption that collapse of the stiffened plate occurs, when any point of the plate reaches the yield stress σ_y, then the Perry-Robertson formula may be used to calculate the load carrying capacity. The following analysis assumes

(S1) $$A_1(t,N,t_1,d) := \sigma_{cr} - \sigma_e \leq 0 \ ,$$

where σ_e is the Euler buckling stress and σ_{cr} is the critical stress of a plate pinned on all sides and loaded in one direction.

The theory of plates gives

$$(*) \qquad \sigma_{cr} = \frac{\pi^2 E t^2}{12(1-\nu^2)b^2}\left(\frac{b}{L}m + \frac{1}{m}\frac{L}{b}\right)^2 ,$$

where $m = \mathrm{INT}(\frac{L}{b}) := \max\{n \in \mathbb{N} \mid n \leq \frac{L}{b}\}$, compare TIMOSHENKO [7]. The Euler buckling stress is given by

$$\sigma_e = \frac{\pi^2 E r^2}{L^2} ,$$

where r is the radius of gyration of the area $A = bt + dt_1$, which can be calculated as follows:

$$r = \sqrt{\frac{I}{A}} ,$$

$$I = \frac{bt^3}{12} + bt\left(y_1 - \frac{t}{2}\right)^2 + \frac{t_1 d^3}{12} + \left(y_1 - t - \frac{d}{2}\right)^2 dt_1 ,$$

$$y_1 = \left[\frac{bt^2}{2} + dt_1\left(\frac{d}{2} + t\right)\right]/A.$$

For stiffened plates such that $b/t \geq 30$ or

$$\text{(S2)} \qquad A_2(t,N,t_1,d) := N - \frac{B}{30t} \leq 0$$

only a part of the plate is effective, when buckling occurs (compare LAWO et al.[4]). In this case one has to calculate the so-called "effective width"

$$b_e := \frac{\sigma_m}{\sigma_y} b$$

and to use a modified Perry-Robertson formula (compare MURRAY [6]). The quotient σ_m/σ_y can be determined from Figure 5 in [6] or can be computed by the following iterative formula as a function of σ_{cr}/σ_y:

$$\frac{\sigma_m}{\sigma_y} = 0.36 + 0.83\left(\frac{\sigma_{cr}}{\sigma_y}\right) - \frac{0.19(\sigma_m/\sigma_y)^2}{(\sigma_{cr}/\sigma_y)}$$

$$- \frac{0.104 \cdot (\sigma_{cr}/\sigma_y)}{\sqrt{\frac{8}{3(1-\nu^2)} (1 - \frac{\sigma_m}{\sigma_y}) + 1.95 \cdot 10^{-2}}} \quad ;$$

this formula is taken from WALKER and DAVIES [9]. Thus, we can calculate the reduced quantities as follows:

$$A' = b_e t + dt_1 \ ,$$

$$y_1' = \left[\frac{b_e t^2}{2} + dt_1 (\frac{d}{2} + t) \right] / A' \ ,$$

$$I' = \frac{b_e t^3}{12} + b_e t (y_1' - \frac{t}{2})^2 + \frac{t_1 d^3}{12} + dt_1 (y_1' - t - \frac{d}{2})^2 \ ,$$

$$r' = \sqrt{I'/A'} \ ,$$

and consequently

$$\sigma_e' = \frac{\pi^2 E (r')^2}{L^2} \ .$$

During the buckling of the plate there is a shift in the position of the neutral axis. This effect must be added to the initial deflection as follows:

$$\Delta' = (y_1' - y_1) \left(1 - \frac{A\sigma_{cr}}{A\sigma_e} \right)$$

and

$$\eta_{'} = \frac{y_1' (a_o + \Delta')}{(r')^2} \ .$$

Using the modified Perry-Robertson formula, we obtain for the axial load at failure

(P3) $$P_f = \sigma_f' A' N \ ,$$

where

$$\sigma_f = \frac{\sigma_y}{2} \left[1 + (1 + \eta') \frac{\sigma_e'}{\sigma_y} \right] - \sigma_y \sqrt{\frac{1}{4} \left[1 + (1 + \eta') \frac{\sigma_e'}{\sigma_y} \right]^2 - \frac{\sigma_e'}{\sigma_y}} \ .$$

The following consideration leads to a further constraint. In fact, stiffener buckling leads to sudden collapse without warning and should be avoided. This can be achieved by making the buckling stress of the stiffener greater than that of the plate. Considering the stiffener to be a plate with three sides pinned and one side free, we obtain the constraint

(S3) $\quad A_3(t,N,t_1,d) := \sigma_{cr} - \sigma_{cr2} \leqq 0,$

where σ_{cr} is defined in equation (*) and where σ_{cr2} denotes the critical stress of a plate free on one side and pinned on the other three sides. This stress is given by

$$\sigma_{cr2} = \frac{\pi^2 E t_1^2 (0.456 + (d/L)^2)}{12(1 - \nu^2)d^2} \quad ,$$

compare TIMOSHENKO [7].

We assume, that at least one stiffener is used. Thus we have the further side condition

(S4) $\quad A_4(t,N,t_1,d) := - N + 1 \leqq 0.$

Since the quantities t, t_1 and d are non-negative and since t_1 is bounded by b, there are the side conditions

(S5) $\quad A_5(t,N,t_1,d) := - t \leqq 0,$

(S6) $\quad A_6(t,N,t_1,d) := - t_1 \leqq 0,$

(S7) $\quad A_7(t,N,t_1,d) := - d \leqq 0,$

(S8) $\quad A_8(t,N,t_1,d) := Nt_1 - B \leqq 0.$

If we introduce the functions

$$p_1(t,N,t_1,d) := \frac{W(t,N,t_1,d)}{\mu L} = (bt + dt_1)N,$$

$$p_2(t,N,t_1,d) := \frac{C(t,N,t_1,d)}{\mu L}$$

$$= (bt + dt_1)kN + Q(t_{min})N/\mu,$$

$$p_3(t,N,t_1,d) := - P_f = - \sigma_f' A' N,$$

compare the equations (P1), (P2), and (P3)), the optimal design problem can be formulated as follows:

<u>VOP.</u> Determine an efficient point of the vector function

$$p(t,N,t_1,d) := (p_1(t,N,t_1,d),\ p_2(t,N,t_1,d),\ p_3(t,N,t_1,d))$$

subject to the sideconditions

$$A_j(t,N,t_1,d) \leqq 0, j = 1,2,\dots,8.$$

We introduce the following dimensionless quantities

$$\tau := t/L,\quad \tau_1 := t_1/L,\quad \delta := d/L,\quad \beta := B/L,$$

$$\beta_e := \frac{\sigma_m}{\sigma_y}\beta,$$

and the variable

$$\alpha_1 := \tau_1\delta.$$

Then we obtain

$$\rho^2 := \left(\frac{r}{L}\right)^2$$

$$= \frac{(\beta\tau^3 + N\alpha_1\delta^2)(\beta\tau + N\alpha_1) + 3\beta\tau N\alpha_1(\tau+\delta)^2}{12(\beta\tau + N\alpha_1)^2}$$

and the sidecondition (S_1) is transformed into

(S_1^*) $$A_1^*(\tau,N,\alpha_1,\delta) := \left(1 + \left(\frac{N}{\beta m}\right)^2\right)^2 \cdot \frac{m^2\tau^2(\beta\tau + N\alpha_1)^2}{1 - \nu^2}$$

$$- (\beta\tau^3 + N\alpha_1\delta^2)(\beta\tau + N\alpha_1) - 3\beta\tau N\alpha_1(\tau + \delta)^2 \leqq 0.$$

The other sideconditions are transformed into

(S_2^*) $$A_2^*(\tau,N,\alpha_1,\delta) := 30N\tau - \beta \leqq 0,$$

(S_3^*) $$A_3^*(\tau,N,\alpha_1,\delta) := \frac{12(1 - \nu^2)(\sigma_{cr} - \sigma_{cr2})}{\pi^2 \cdot E}$$

$$= \tau^2 m^2 (1 + (\frac{N}{\beta m})^2)^2 - \frac{(0.456 + \delta^2)\alpha_1^2}{\delta^4} \leqq 0 ,$$

(S_4^*) $\qquad A_4^*(\tau,N,\alpha_1,\delta) := - N + 1 \leqq 0 ,$

(S_5^*) $\qquad A_5^*(\tau,N,\alpha_1,\delta) := - \tau \leqq 0 ,$

(S_6^*) $\qquad A_6^*(\tau,N,\alpha_1,\delta) := - \alpha_1 \leqq 0 ,$

(S_7^*) $\qquad A_7^*(\tau,N,\alpha_1,\delta) := - \delta \leqq 0 ,$

(S_8^*) $\qquad A_8^*(\tau,N,\alpha_1,\delta) := N\delta_1 - \beta\delta \leqq 0.$

The objective functions are transformed into

$$P_1^*(\tau,N,\alpha_1,\delta) := \beta\tau + \alpha_1 N ,$$

$$P_2^*(\tau,N,\alpha_1,\delta) := (\beta\tau + \alpha_1 N) \cdot k + Q(\tau_{min})N/\mu ,$$

where

$$\tau_{min} := \min(\tau,\alpha_1/\delta) ,$$

and

$$P_3^*(\tau,N,\alpha_1,\delta) := P_3(\tau,N,\alpha_1,\delta)/\sigma_y$$

$$= \frac{1}{2}\Gamma - \sqrt{\frac{1}{4}\Gamma^2 - \xi\rho'^2} ,$$

where

$$\xi := \pi^2 E/\sigma_y ,$$

$$\rho'^2 := \frac{(\beta_e\tau^3 + N\alpha_1\delta^2)(\beta_e\tau + N\alpha_1) + 3\beta_e\tau N\alpha_1(\tau+\delta)^2}{12(\beta_e\tau + N\alpha_1)^2}$$

and

$$\Gamma := 1 + (1 + \tilde{\eta}')\xi \cdot \rho'^2 .$$

Here $\tilde{\eta}'$ denote the quantity η' as a function of the new variables τ,α_1,δ and N.

Thus we obtain the transformed vector optimization problem

<u>VOP*.</u> Determine an efficient point of the vectorfunction $p^*(\tau,N,\alpha_1,\delta)$ subject to the sideconditions

$$A_j(\tau,N,\alpha_1,\delta) \leqq 0,\ j = 1,2,\ldots,8.$$

4. The structure of the feasible set.

For each $\tau \geqq 0$ and each $N \geqq 1$ we introduce the set

$$Z_{\tau,N} := \bigcap_{j=1}^{8} \{(\alpha_1,\delta) \in \mathbb{R}^2 \mid A_j^*(\tau,N,\alpha_1,\delta) \leqq 0\}.$$

The inequalities $(S_1^*),(S_2^*),\dots,(S_8^*)$ imply, that $Z_{\tau,N}$ is given by

$$Z_{\tau,N} = \bigcap_{j=1,3,8} \{(\alpha_1,\delta) \in \mathbb{R}_+^2 \mid A_j^*(\tau,N,\alpha_1,\delta) \leqq 0\}$$

where

$$\mathbb{R}_+^2 = \{(\alpha_1,\delta) \in \mathbb{R}^2 \mid \alpha_1 \geqq 0 \ \&\ \delta \geqq 0\}.$$

The inequality $A_1^*(\tau,N,\alpha_1,\delta)$ is equivalent to

$$P(\lambda) := \lambda^2 K + \lambda H + M \geqq 0,$$

where we have used the abbreviations

$$\lambda := \delta\sqrt{\alpha_1},$$

$$K := (4\beta\tau + N\alpha_1)N,$$

$$H := 6\beta\tau\sqrt{\alpha_1}\, N\,,$$

$$M := \beta\tau^3[\beta\tau + 4\alpha_1 N] - \frac{\omega^2(\beta\tau + \alpha_1 N)^2}{1-\nu^2}\,,$$

and

$$\omega^2 := \left(1 + \left(\frac{N}{\beta m}\right)^2\right)^2 \tau^2 m^2 .$$

An easy computation shows that

$$P(\lambda) \geqq 0 \quad \& \quad \lambda \geqq 0$$

is equivalent to

$$\delta \geqq \delta_1(\alpha_1) := \begin{cases} 0 & \text{if } M - \frac{H^2}{2K} \geqq 0 \\ \max\left[0, \left(-\frac{H}{K} + \sqrt{\left(\frac{H}{2K}\right)^2 - \frac{M}{K}}\right) / \sqrt{\alpha_1}\right] & \text{if } M - \frac{H^2}{2K} < 0. \end{cases}$$

The inequalities $A_3(\tau,N,\alpha_1,\delta) \leqq 0$ and $A_8(\tau,N,\alpha_1,\delta) \leqq 0$ equivalent to

$$\delta \leqq \delta_3(\alpha_1) := \frac{\alpha_1^2 + \sqrt{\alpha_1^4 + 1{,}824\omega^2\alpha_1^2}}{2\omega^2}$$

and

$$\delta \leqq \delta_8(\alpha_1) := \frac{N\alpha_1}{\beta} ,$$

respectively. Thus, we obtain

$$Z_{\tau,N} = \{(\alpha_1,\delta) \in \mathbb{R}_+^2 \,|\, \delta_0(\alpha_1) \leqq \delta \leqq \delta_3(\alpha_1)\}$$

where $\delta_0(\alpha_1) := \max(\delta_1(\alpha_1), \delta_8(\alpha_1))$.

If only the thickness τ of the plate is given, then we obtain

$$Z_\tau := \bigcap_{j=1}^{8} \{(\alpha_1,\delta,N) \in \mathbb{R}^2 \times \mathbb{N} \,|\, A_j^*(\tau,N,\alpha_1,\delta) \leqq 0\}$$

$$= \bigcup_{N=1}^{N_m} Z_{\tau,N} \times \{N\} ,$$

where N_m is the maximal possible number of stiffeners. This number is given by

$$N_m := \mathrm{INT}(\beta/30\tau) ,$$

compare inequality (S_2^*).

Now assume that the thickness τ of the plate is also a design variable. To avoid unrealistic considerations for $\tau \longrightarrow 0$, we assume that τ satisfies the inequality

$$(S_9^*) \qquad A_9^*(\tau,N,\alpha_1,\delta) := \gamma_0 - \tau \leqq 0,$$

where $\gamma_0 > 0$ is a given constant: For each $N \geqq 1$ we introduce the set

$$Z_N := \bigcap_{j=1}^{9} \{(\alpha_1,\delta,\tau) \in \mathbb{R}^3 \,|\, A_j^*(\tau,N,\alpha_1,\delta) \leqq 0\}.$$

Then we have

$$Z_N = \bigcup_{\tau \geqq \gamma_0} Z_{\tau,N} \times \{\tau\} ,$$

and the feasible set can be written as

$$Z = \bigcup_{N \in \mathbb{N}} Z_N \times \{N\}.$$

Of course, this union contains only a finite number of non-empty expressions $Z_N \times \{N\}$. In fact, for N large enough we have

$$\frac{30\beta}{N} < \gamma_o .$$

Then the side conditions

$$\tau \leqq \frac{30\beta}{N} \quad \& \quad \gamma_o \leqq \tau$$

imply that Z_N is empty.

5. Given thickness of the plate.

In this section we consider the optimal design problem in the case of weight minimization, maximization of the collapsing load, and with given thickness of the plate. For $N = 1,2,...,N_m$ we define the functions

$$q_1^N(\alpha_1,\delta) := p_1(\tau,N,\alpha_1,\delta), \; q_2^N := p_3(\tau,N,\alpha_1,\delta) \; ,$$

$$q^N := (q_1^N, q_2^N).$$

Then the optimal design problem can be stated as

VOP. Determine all q-efficient points in the set

$$Z_\tau = \bigcup_{N=1}^{N_m} Z_{\tau,N} \times \{N\}.$$

To solve this vector optimization problem, we consider for $N = 1,2,...,N_m$ the vector optimization problem

VOP(N). Determine all q_N-efficient points in the

$$Z_{\tau,N} := \{(\alpha_1,\delta) \in \mathbb{R}^2_+ | \; \delta_o^N(\alpha_1) \leqq \delta \leqq \delta_3(\alpha_1)\}.$$

Following the approach in [2], we introduce for each $\lambda \in \mathbb{R}$ the minimization problem

MP(N,λ). Minimize the objective function

$$Q(\alpha_1,\delta,z) := z$$

subject to the side-conditions

$$q_1^N(\alpha_1,\delta) - \lambda = \beta\tau + \alpha_1 N - \lambda \leqq z,$$

$$q_2^N(\alpha_2,\delta) \leqq z,$$

$$\delta_o^N(\alpha_1) \leqq \delta \leqq \delta_3^N(\alpha_1),$$

$$\delta \geqq 0, \; \alpha_1 \geqq 0.$$

We denote by $P_N(\lambda)$ the set of all minimal points of $MP(N,\lambda)$.

The parametric minimization problem $MP(N,\lambda)$ is related with the vector optimization problem VOP(N) by the following theorems (compare[2]):

<u>Theorem 5.1.</u> Let $x_o = (\alpha_1^o, \delta_o)$ be a q_N-efficient point of the problem VOP(N). Then there exist a parameter $\lambda \in \mathbb{R}$ and a real number z_o such that (x_o, z_o) is a minimal point of $MP(N,\lambda)$.

<u>Theorem 5.2.</u> If $Z_{\tau,N}$ is non-empty, then for each $\lambda \in \mathbb{R}$ the set $P_N(\lambda)$ contains at least one element (x_o, z_o) such that x_o is a q_N-efficient point.

<u>Proof:</u> The assumption $Z_{\tau,N} \neq \emptyset$ implies $Z^* \neq \emptyset$, where Z^* denotes the feasible set of the minimization problem $MP(N,\lambda)$. Since $z \geq \beta\tau - \lambda$, it follows that also the minimum value E^* satisfies the inequality $E^* \geq \beta\tau - \lambda$. Thus it suffices to consider only those z with $\beta\tau - \lambda \leq z \leq 2E^*$. The last inequality implies

$$0 \leq \alpha_1 \leq \frac{2E^* + \lambda - \beta\tau}{N} .$$

Thus, it suffices to minimize the objective function $Q(\alpha_1, \delta, z)$ on the compact set

$$Z^* \cap \{(\alpha_1, \delta, z) \in \mathbb{R}_+^2 \times \mathbb{R} \mid \beta\tau - \lambda \leq z \leq 2E^*\}$$

$$\cap \{(\alpha_1, \delta, z) \in \mathbb{R}_+^2 \times \mathbb{R} \mid \delta_o^N(\alpha_1) \leq \delta \leq \delta_3^N(\alpha_1) \;\&\; 0 \leq \alpha_1 \leq \frac{2E^* + \lambda - \beta\tau}{N}\}.$$

It follows that $P_N(\lambda)$ is non-empty and compact. Now apply theorem 2.2 of [2].

□

We prove the following refinements of these theorems:

<u>Theorem 5.3.</u> Let $x_o = (N_o, \alpha_1^o, \delta_o)$ be a q-efficient point of the problem VOP. Then (α_1^o, δ_o) is a q_{N_o}-efficient point

of the problem VOP(N_o) and there exist a parameter $\lambda \in \mathbb{R}$ and a real number z_o such that (x_o, z_o) is a minimal point of MP(N_o, λ).

Proof: It is easy to see, that (α_1^o, δ_o) is a q_{N_o}-efficient point of VOP(N_o). Then the result follows from Theorem 5.1.

□

Theorem 5.4. If Z_τ is non-empty, then for each $\lambda \in \mathbb{R}$ there exists a natural number N, $1 \leq N \leq N_m$, and an element (x_N, z_N) in $P_N(\lambda)$ such that x_N is a q-efficient point.

Proof: The assumption $Z_\tau \neq \emptyset$ implies that the set

$$\Psi := \{N \in \mathbb{N} \mid Z_{\tau,N} \neq \emptyset\}$$

is also non-empty. By theorem 5.2, for each $N \in \Psi$ there exists an element $(x_N, z_N) \in P_N(\lambda)$ such that x_N is q_N-efficient. Define

$$z_o := \min_{N \in \Psi} z_N$$

and the set

$$\Psi_o := \{N \in \Psi \mid z_N = z_o\}.$$

Each $x_N, N \in \Psi_o$, satisfies the inequalities

$$q_1^N(x_N) \leq \lambda + z_N \quad \& \quad q_2^N(x_N) \leq z_N ,$$

and at least one inequality is active. Thus one of the sets

$$\Psi_1 := \{N \in \Psi_o \mid q_1^N(x_N) = \min_{n \in \Psi_o} q_1^n(x_n)\}$$

and

$$\Psi_2 := \{N \in \Psi_o \mid q_2^N(x_N) = \min_{n \in \Psi_o} q_2^n(x_n)\}$$

is non-empty. Then we have one of the following cases:

(1) If $q_1^N(x_N) < z_o + \lambda$ for $N \in \Psi_1$, then x_N is q-efficient ;

(2) If $q_2^N(x_N) < z_o$ for $N \in \Psi_2$, then x_N is q-efficient ;

(3) If $q_1^N(x_N) = z_o + \lambda$ for $N \in \Psi_1$ and $q_2^N(x_N) = z_o$ for $N \in \Psi_2$, then each point in Ψ_o is q-efficient.

□

By theorem 5.3 the optimal design problem is reduced to N_m vector optimization problems VOP(N), N = 1,2,..,N_m. Each of these vector optimization problems can be solved numerically by the method developed in [2], i.e. one has to solve the parametric minimization problems MP(N,λ), $\lambda \in \mathbb{R}$. Assume we have solved the problems MP(N,λ)N = 1,2,...,N_m for a given $\lambda \in \mathbb{R}$. Then the proof of theorem 5.4 shows how we can calculate a point

$$(x_o, z_o) \in P(\lambda) := \bigcup_{N=1}^{N_m} P_N(\lambda)$$

such that x_o is a q-efficient point. This leads to the following algorithm, which calculates for a given $\lambda \in \mathbb{R}$ a point $(x_o, \lambda_o) \in P(\lambda)$ such that x_o is q-efficient.

<u>Algorithm</u>

STEP 1. $\Psi := \emptyset$;

STEP 2. <u>FOR</u> N = 1,2,...,N_m <u>DO</u>

BEGIN

STEP 3. Calculate a solution (x_N^o, z_N) of MP(N,λ) ;

STEP 4. Calculate the numbers

$$\eta_1 := \lambda + z_N \quad \& \quad \eta_2 := z_N \ ;$$

STEP 5. Calculate solutions v_1^N resp. v_2^N of the minimum problems

<u>HP1.</u> Minimize $q_1^N(x)$ subject to $x \in Z_{\tau,N}$ and $q_2^N(x) = \eta_2$,

resp.

<u>HP2.</u> Minimize $q_2^N(x)$ subject to $x \in Z_{\tau,N}$ and $q_1^N(x) = \eta_1$;

STEP 6. (a) If $q_1^N(v_1^N) < \eta_1$, then v_1^N is q_N-efficient ,

(b) If $q_2^N(v_2^N) < \eta_2$, then v_2^N is q_N-efficient ,

(c) If $q_1^N(v_1^N) = \eta_1$ and $q_2^N(v_2^N) = \eta_2$,

then v_1^N and v_2^N are q_N-efficient ;

STEP 7. If (a),(b) or (c) is true, then

$\Psi := \Psi \cup \{N\}$;

END

STEP 8. Calculate $z_o := \min_{N\in\Psi} z_N$ and the sets Ψ_o, Ψ_1, Ψ_2 (compare the proof of theorem 5.4).

STEP 9. (a) If $q_1^N(v_1^N) < z_o + \lambda$ for $N \in \Psi_1$, then v_1^N is q-efficient,

(b) If $q_2^N(v_2^N) < z_o$ for $N \in \Psi_2$, then v_2^N is q-efficient,

(c) If $q_1^N(v_1^N = z_o + \lambda$ for $N \in \Psi_1$, and $q_2^N(v_2^N = z_o$ for $N \in \Psi_2$, then each of the points v_1^N and v_2^N is q-efficient. □

The set

$$B_q := \{q(x) \in \mathbb{R}^2 \,|\, x \text{ is q-efficient}\}$$

is called the q-efficient boundary of the transformed feasible set $q(Z_\tau)$. For the calculation of B_q (with precision $\varepsilon > 0$), we can proced as follows:

(1) Determine a $\lambda_o \in \mathbb{R}$ and a solution (x_o,z_o) in $P(\lambda_o)$ such that

(*) $\forall_{\lambda\leq\lambda_o} \quad (x_o,z_\lambda) \in P(\lambda)$

and x_o is q-efficient ;

(2) Determine for $\lambda_j = \lambda_o + j\varepsilon$, $j = 1,2,\ldots,$ a q-efficient point in $P(\lambda_j)$.

To prove the existence of such a λ_o, we use the representation of the feasible set

$$Z_\tau = \bigcup_{N=1}^{N_m} \{(\alpha_1,\delta) \in \mathbb{R}_+^2 \,|\, \delta_o^N(\alpha_1) \leq \delta \leq \delta_3^N(\alpha_1)\}.$$

For each $1 \leq N \leq N_m$ with $Z_{\tau,N} \neq \emptyset$, there exists exactly one $(\alpha_1(N),\delta(N)) \in Z_{\tau,N}$ such that $\alpha_1(N)$ is as small as possible. Since $q_1^N((\alpha_1,\delta) = \tau\beta + \alpha_1 N$ is linear in α_1 and independent of δ, it attains its infimum on $Z_{\tau,N}$ in the point $(\alpha_1(N),\delta(N))$. Define the quantities

$$u_1^N := q_1^N(\alpha_1(N),\delta(N)) \quad ,$$

$$\Gamma_o := \{N \in \mathbb{N} \,|\, u_1^n = \min_{Z_{\tau,n}\neq\emptyset} u_1^n\} \quad ,$$

$$\Gamma^* := \{N^* \in \mathbb{N} \mid q_2^{N^*}(\alpha_1(N^*),\delta(N^*)) = \min_{n\in\Gamma_o} q_2^n(\alpha_1(n),\delta(n))\}.$$

We introduce the abbreviations

$$\alpha_1^* := \alpha_1(N^*) \quad \& \quad \delta^* := \delta(N^*), \quad N^* \in \Gamma^*.$$

Then

$$\lambda_o := q_1^{N^*}(\alpha_1^*,\delta^*) - q_2^{N^*}(\alpha_1^*,\delta^*) ,$$

$$x_o := (\alpha_1^*,\delta^*,N^*) ,$$

$$z_o := q_2^{N^*}(\alpha_1^*,\delta^*)$$

have the required properties.

In fact, if (x_o,z_o) is not a minimal point of $MP_{N^*}(\lambda_o)$, then there exist a point (α_1,δ) in Z_{τ,N^*} and a real number $z < z_o$ such that

$$q_1^{N^*}(\alpha_1,\delta) - \lambda_o \leqq z < z_o = q_1^{N^*}(\alpha_1^*,\delta^*) - \lambda_o ,$$

or

$$q_1^{N^*}(\alpha_1,\delta) < q_1^{N^*}(\alpha_1^*,\delta^*) ,$$

which is impossible.

To prove the q-efficiency of the point x_o, choose an element $x = (\alpha,\delta,N)$ in Z_τ such that

$$q_1^N(\alpha,\delta) \leqq q_1^{N^*}(\alpha_1^*,\delta^*)$$

and

$$q_2^N(\alpha,\delta) \leqq q_2^{N^*}(\alpha_1^*,\delta^*).$$

By definition of x_o we have

$$q_1^{N^*}(\alpha_1^*,\delta^*) \leqq q_1^N(\alpha_1,\delta)$$

and consequently

$$q_1^{N^*}(\alpha_1^*,\delta^*) = q_1^N(\alpha_1,\delta).$$

Using again the definition of x_o we have

$$q_2^{N^*}(\alpha_1^*,\delta^*) \leqq q_2^N(\alpha_1,\delta)$$

and consequently

$$q_2^{N^*}(\alpha_1^*,\delta^*) = q_2^N(\alpha_1,\delta) ,$$

i.e. x_o is q-efficient.

To prove the relation (*), choose a $\lambda < \lambda_o$ and let $x = (\alpha_1, \delta, N)$ be a q-efficient point contained in $P(\lambda)$. By definition of x_o we have

$$q_1^{N^*}(\alpha_1^*, \delta^*) \leqq q_1^N(\alpha_1, \delta) ,$$

which implies

$$q_1^{N^*}(\alpha_1^*, \delta^*) - \lambda_o < q_1^{N^*}(\alpha_1, \delta) - \lambda$$

$$\leqq q_1^N(\alpha_1, \delta) - \lambda \leqq z_\lambda .$$

Then we have

$$q_2^{N^*}(\alpha_1^*, \delta^*) = q_1^{N^*}(\alpha_1^*, \delta^*) - \lambda_o \leqq z_\lambda ,$$

i.e. the element (x_o, z_λ) is contained in $P(\lambda)$. □

Using the special structure of the feasible set $Z_{\tau,N}$ and the objective functions q_1^N and q_2^N we obtain the following simplifications for the calculation of the solutions of the minimization problems (HP1) and (HP2) of the algorithm. In fact, we have for the minimization problem (HP1):

<u>(HP1a).</u> Minimize $q_1^N(\alpha_1, \delta) = \beta\tau + \alpha_1 N$

subject to

(a) $q_2^N(\alpha_1, \delta) = \eta_2$

and

$$\delta_o^N(\alpha_1) \leqq \delta \leqq \delta_3^N(\alpha_1).$$

From side condition (a) we can determine δ as a function of α_1, say $\delta = \overline{\delta}_N(\alpha_1)$. Thus, the minimization problem (HP1) reduces to the calculation of the least α_1 such that

$$\delta_o^N(\alpha_1) \leqq \overline{\delta}_N(\alpha_1) \leqq \overline{\delta}_3(\alpha_1).$$

The special form of the functions δ_o^N, $\overline{\delta}_N$ and δ_3^N imply that this is equivalent to calculate the less zero of the least zeros of

$$\delta_o^N(\alpha_1) - \overline{\delta}_N(\alpha_1) \quad \text{resp.} \quad \overline{\delta}_N(\alpha_1) - \delta_3^N(\alpha_1).$$

In the case of the minimization problem (HP2) we remark that q_1^N is independent of δ and linear in α_1. Thus (HP2) is equivalent to

<u>(HP2a).</u> Minimize $H_N(\delta) := q_2^N(\frac{\eta_1 - \beta\tau}{N}, \delta)$

subject to the sidecondition

$$\delta_o(\frac{\eta_1 - \beta\tau}{N}) \leqq \delta \leqq \delta_3(\frac{\eta_1 - \beta\tau}{N}).$$

These simplifications will be lost, if we consider cost minimization instead of weight minimization, since the costs depend not only on α_1 but also on δ.

6. Variable thickness of the plate

These case of a variable thickness can be treated in a similar way like the case of a given thickness. For $n \in \mathbb{N}$, we define the functions

$$q_1^N(\alpha_1,\delta,\tau) := p_1(\tau,N,\alpha_1,\delta),\quad q_2^N(\alpha_1,\delta,\tau) := p_3(\tau,N,\alpha_1,\delta),$$

$$q_N := (q_1^N,\ q_2^N).$$

Then the optimal design problem can be stated as

<u>VOP.</u> Determine all q-efficient points in the set

$$Z = \bigcup_{N\in\mathbb{N}} Z_N \times \{N\}.$$

To solve this vector optimization problem we consider for $N \in \mathbb{N}$ the vector optimization problem

<u>VOP(N).</u> Determine all q_N-efficient points in the set Z_N.

Like in section 5 we consider for each $\lambda \in \mathbb{R}$ the minimization problem

<u>MP(N,λ).</u> Minimize the objective function

$$Q(\alpha_1,\delta,\tau,z) := z$$

subject to the side condition

$$q_1^N(\alpha_1,\delta,\tau) - \lambda = \beta\tau + N\alpha_1 - \lambda \leqq z,$$

$$q_2^N(\alpha_1,\delta,\tau) \leqq z,$$

$$\delta_o^N(\alpha_1,\tau) \leqq \delta \leqq \delta_3^N(\alpha_1,\tau),$$

$$0 \leqq \delta,\ 0 \leqq \alpha_1,\ \gamma_o \leqq \tau.$$

We denote by $P_N(\lambda)$ the set of all minimal points of MP(N,λ). Further we introduce the set

$$P(\lambda) := \bigcup_{N\in\mathbb{N}} P_N(\lambda).$$

The following theorems show how the parametric minimization problem MP(N,λ) is related with the vector optimization problem VOP. The proofs are with slight modifications the same as in section 5 and are omitted.

Theorem 6.1. Let $(\alpha_1^o, \delta_o, \tau_o, N_o)$ be a q-efficient point of the problem VOP. Then $(\alpha_1^o, \delta_o, \tau_o)$ is a q_{N_o}-efficient point of the problem VOP(N_o) and there exist numbers $\lambda, z_o \in \mathbb{R}$ such that

$$((\alpha_1^o, \delta_o, \tau_o), z_o)$$

is a minimal point of MP(N_o, λ).

Theorem 6.2. If Z is non-empty, then for each $\lambda \in \mathbb{R}$ there exist $N \in \mathbb{N}$ and $(x_N, z_N) \in P_N(\lambda)$ such that (x_N, N) is q-efficient.

For the calculation of q-efficient points in $P_N(\lambda)$ the same algorithm as in section 5 can be used. Also for the calculation of the q-efficient boundary B_q (with precision $\varepsilon > 0$), we can proceed like in section 5:

(1) Determine a $\lambda_o \in \mathbb{R}$, a solution (x_o, z_o) in $P(\lambda_o)$, and an $N \in \mathbb{N}$ such that

$$(*) \quad \forall_{\lambda \leq \lambda_o} \quad (x_o, z_\lambda) \in P(\lambda)$$

and (x_o, N) is q-efficient ;

(2) Determine for $\lambda_j = \lambda_o + j\varepsilon$, $j = 1,2,\ldots$, a q-efficient point in $P(\lambda_j)$.

To prove the existence of such a λ_o, we use the representation of the sets

$$Z = \bigcup_{N \in \mathbb{N}} Z_N \times \{N\}$$

and

$$Z_N = \bigcup_{\tau \geq \gamma_o} Z_{\tau,N} \times \{\tau\},$$

compare section 4. For each N with $Z_N \neq \emptyset$ let $(\alpha_1^N, \delta_N, \tau_N)$ be a minimal point of q_1^N in Z_N. Define the quantities

$$u_1^N := q_1^N(\alpha_1^N, \delta_N, \tau_N) \quad ,$$

$$\Gamma_o := \{N \in \mathbb{N} \mid u_1^N = \min_{Z_{\tau,n} \neq \emptyset} u_1^n\}$$

$$\Gamma^* := \{N^* \in \mathbb{N} \mid q_2^{N^*}(\alpha_1(N^*), \delta(N^*), \tau(N^*)) = \min_{n \in \Gamma_o} q_2^n(\alpha_1(n), \delta(n), \tau(n))$$

and proceed like in section 5. We remark that there are at most finitely many $N \in \mathbb{N}$ with $Z_N \neq \emptyset$.

Using the special structure of the feasible set Z_N and of the objective function q_1^N we obtain the following simplification for the calculation of a minimum point of (HP2). In fact, we obtain for the minimization problem (HP2):

(HP2a). Minimize

$$H_N(\delta,\tau) := q_2^N\left(\frac{\eta_1 - \beta\tau}{N}, \delta, \tau\right)$$

subject to the side conditions

$$\delta_o\left(\frac{\eta_1 - \beta\tau}{N}, \tau\right) \leqq \delta \leqq \delta_3\left(\frac{\eta_1 - \beta\tau}{N}, \tau\right)$$

and

$$O \leqq \delta, \quad O \leqq \alpha_1, \quad \gamma_o \leqq \tau \leqq \beta/N.$$

This simplification will be lost, if consider cost minimization instead of weight minimization, since the costs depend not only on α_1 and τ but also on δ.

References

[1] Brosowski,B.; Conci,A. (1983) On the optimal design of stiffened plates. Anais VII. Congresso Brasileiro de Engenharia Mecânica, Vol.D, 169-170. Uberlandia (Brasil).

[2] Brosowski,B.; Conci,A. (1983) On vector optimization and parametric programming. Segundas Jornadas Latino Americanas de Matematica Aplicada, Vol.2, 483-495. Rio de Janeiro (Brasil).

[3] Guddat,J. (1979) Parametrische Optimierung und Vektoroptimierung. In: Lommatzsch [5], 54-75.

[4] Lawo,M., Murray,N.W., Thierauf,G. (1978) Optimization of stiffened plates. Civil Engineering Transactions. 13-16.

[5] Lommatzsch,K.(ed.) (1979) Anwendungen der linearen parametrischen Optimierung. Birkhäuser-Verlag, Basel,Stuttgart.

[6] Murray,N.W. (1975) Analysis and design of stiffened plates for collapse load. The structural Engineer. 53, 153-158.

[7] Timoshenko,S.P. (1961) Theory of elestic stability. McGraw-Hill, London.

[8] Vester,H. (1973) Automatische Kostenkalkulation von maßgeschneiderten Stahlgeschoßtragwerken. Diplomarbeit, Ruhr-Universität Bochum, Institut für Konstr.Ingenieurbau.

[9] Walker,A.C., Davies,P. An elementary study on non-linear buckling phenomena in stiffened plates.

[10] Zeleny,M. (1970) Linear multiobjective programming. Lecture Notes in Economics and Mathematical Systems, Vol.95. Springer-Verlag Berlin, Heidelberg, New York.

Ackowledgement:

I thank Aura Conci and Khosrow Ghavami for many helpful discussions.

Address:

Prof.Dr.Bruno Brosowski, Fachbereich Mathematik der Johann Wolfgang Goethe-Universität Frankfurt, Robert-Mayer-Straße 6-10, D 6000 Frankfurt(M), West Germany.

International Series of
Numerical Mathematics, Vol. 72

BEST APPROXIMATION BY SMOOTH FUNCTIONS AND RELATED PROBLEMS

A.L. Brown
School of Mathematics, University of Newcastle upon Tyne

1. Introduction

SATTES (1980) has considered best uniform approximation by smooth functions in the space $C([-1,1])$ of continuous real valued functions on an interval and has characterised such best approximations. GLASHOFF (1980) and PINKUS (1980) have considered a family of similar problems. We state a general problem which contains those of Sattes, Glashoff and Pinkus as special cases. A general theorem which characterises best approximations is proved in Section 2 and in Section 4 it is used to obtain an alternative proof of Sattes' result. The link between the general theorem and the proof of Sattes' result is provided by a theorem, concerning the zeros of certain functions, which is proved in Section 3. Section 3 is a small contribution to the theory of total positivity.

Let X be a compact Hausdorff topological space and λ_0 a regular Borel measure on X. There are two particular cases of interest: those in which X is either an interval or a circle and λ_0 is Lebesgue measure. Let $C(X)$, $L^1(X,\lambda_0)$ and $L^\infty(X,\lambda_0)$ be the usual spaces of real valued functions equipped with their usual norms. Let $K: X\times X\to \mathbb{R}$ be a kernel defining an integral operator

$$T_K : L^\infty(X,\lambda_0) \to C(X)$$

by the equation

$$(T_K f)(x) = \int K(x,y)f(y)d\lambda_0(y).$$

Let M and N be finite dimensional subspaces of the space $C(X)$. Let $(L^\infty)_1$ denote the unit ball $\{f\in L^\infty : \|f\| \le 1\}$ of the space $L^\infty = L^\infty(X,\lambda_0)$, and let

$$N^{\perp} = \{f \in L^1(X,\lambda_0): \int fg\,d\lambda_0 = 0 \text{ for all } g \in N\}.$$

We will write

$$W(K;M,N) = M + T_K((L^{\infty})_1 \cap N^{\perp}).$$

It will be assumed that the kernel $K: X \times X \to \mathbb{R}$ is a Borel measurable function such that $y \to K(\cdot,y)$ defines a continuous mapping of X into the space $L^1(X,\lambda_0)$. In this situation it can be shown by a standard functional analytic argument that $T_K((L^{\infty})_1 \cap N^{\perp})$ is a compact subset of $C(X)$ and, therefore, for each function $\phi \in C(X)$ there exists a best uniform approximation to ϕ from the set $W(K;M,N)$. The General Problem is to characterise such best approximations. We will describe the special cases which are of interest.

SAATES obtained a characterisation of best approximations by smooth functions - functions with rth derivative bounded by one, where $r \geq 2$. (The case $r = 1$ had been considered much earlier by Korneichuk). Let $X = [0,1]$, let λ_0 be Lebesque measure, let $N = \{0\}$ and let $M = P_{r-1}$, the space of polynomial functions of degree $\leq r-1$. Let

$$K_r(s,t) = \frac{(s-t)_+^{r-1}}{(r-1)!} .$$

Then $W(K_r;P_{r-1},\{0\})$ is the set of smooth functions

$$W_{\infty,r} = \{f \in C([0,1]): \quad f^{(r-1)} \text{ abs.conts.}, \|f^{(r)}\|_{\infty} \leq 1\}.$$

Sattes' result includes the fact that for each $\phi \in C([0,1])$ there is a subinterval of $[0,1]$ on which each best approximation to ϕ from $W_{\infty,r}$ is a uniquely determined perfect spline. His characterisation of best approximations is in terms of this property.

The kernel K_r is totally positive (see KARLIN) and certain elements of Sattes' discussion are of a kind familiar to students of total positivity. This circumstance led PINKUS to formulate and consider the next problem. It was considered independently by GLASHOFF for whom it appears to have originated in a control problem for parabolic equations discussed by KARAFIAT.

Let X and λ_0 again be the unit interval $[0,1]$ and Lebesgue measure, let $M = N = \{0\}$ and let K be a strictly totally positive (or slightly more generally, a strictly sign regular) kernel. Such kernels arise in certain boundary value problems for partial differential equations. In this situation

$$W(K;\{0\},\{0\}) = \{\int K(.,t)f(t)dt : \ f \in L^{\infty}([0,1]), \|f\| \leq 1\}.$$

The problem of best approximation to $\phi \in C([0,1])$ from $W(K;\{0\},\{0\})$ can also be regarded as a control problem. Pinkus and Glashoff proved independently that for each $\phi \in C([0,1])$ there exists a unique best approximation from $W(K;\{0\},\{0\})$ and that it is a generalised perfect spline, that is of the form $T_K f$ where f is a step function of constant modulus one. Glashoff's proof combines a functional analytic argument with a standard result from the theory of total positivity. Pinkus' argument is different: it also uses the theory of total positivity but otherwise it is closer in character to that of Sattes.

The general theorem which we prove depends upon an extension of Glashoff's simple functional analytic argument. It transforms the general problem of best approximation to one involving consideration of the zero sets of functions of the form

$$\mu(y) = g(y) + \int K(x,y)\, d\lambda(x) ,$$

where λ is some measure on X, $g \in N$ and λ and g satisfy certain conditions.

In the situations considered by Sattes, Pinkus and Glashoff the subspace N of the general problem is trivial: $N = \{0\}$. The problem has been formulated in a generality which includes the periodic analogue of Sattes' problem. Let $\tilde{C}(\mathbb{R})$ be the space of continuous 2π-periodic functions defined on $\mathbb{R}$. Then $\tilde{C}(\mathbb{R})$ can be identified with C(X), where X is the circle. The set of periodic smooth functions

$$\tilde{W}_{\infty,r} = \{f \in \tilde{C}(\mathbb{R}) : \ f^{(r-1)} \text{abs.conts}, \ \|f^{(r)}\|_{\infty} \le 1\}$$

is essentially of the form $W(D_r : P_0, P_0)$, where D_r is a certain convolution kernel and $P_0 = M = N$ is the one-dimensional space of constant functions. The problem of best approximation to $\phi \in \tilde{C}(\mathbb{R})$ from $\tilde{W}_{\infty,r}$ will be discussed in a later paper.

2. The General Theorem

Let X, λ_0, K, T_K, M and N, and $W = W(K;M,N)$ be as in the statement of the general problem. The set W is a closed convex subset of C(X). If $\phi \in C(X) \setminus W$ then $d(\phi,W)$ will denote the distance, with respect to the uniform norm, of ϕ from W, and

$$B(\phi, d(\phi,W)) = \{f \in C(X): \ \|\phi - f\| < d(\phi,W)\}$$

is the open ball centre ϕ, of radius $d(\phi,W)$. Then $B(\phi,d(\phi,W))$ is an open convex set disjoint from W. Therefore $B(\phi,d(\phi,W))$ and W can be separated by a linear functional $\lambda \in C(X)^*$, the dual space of C(X). We will identify $C(X)^*$ and the space of regular Borel measures on X and will regard $\lambda \in C(X)^*$ interchangeably as a linear functional on C(X) or as a measure on X.

The general theorem which follows characterises a situation involving both a best approximation u_0 in W to $\phi \in C(X)\setminus W$ and a separating functional (measure) λ. In the statement of the theorem $\|.\|_1$ denotes the norm of $L^1 = L^1(X,\lambda_0)$.

Theorem 1. Let $\phi \in C(X)\setminus W$, $\lambda \in C(X)^*\setminus\{0\}$ and $g \in N$; let μ be defined by

$$\mu(y) = g(y) + \int K(x,y)\,d\lambda(x),$$

and let $u_0 = k_0 + T_K f_0 \in W$ where $k_0 \in M$, $f_0 \in L^\infty(X,\lambda_0) \cap N^\perp$ and $\|f_0\| \le 1$. Then the three conditions

I(a) u_0 is a best approximation to ϕ from W,

I(b) $\lambda(u) < \lambda(v)$ for all $u \in W$ and $v \in B(\phi,d(\phi,W))$,

I(c) $\|\mu\|_1 \le \|\mu-h\|_1$ for all $h \in N$, (that is $-g$ is a best L^1-approximation to $\int K(x,\cdot)d\lambda(x)$ from N),

are together equivalent to the three conditions

II(i) $\lambda(k) = 0$ for all $k \in M$,

II(ii) $f_0(y) = \operatorname{sgn}\mu(y)$ for λ_0-almost every y in $X\setminus\mu^{-1}(0)$,

III(iii) $\lambda(\phi - u_0) = \|\lambda\|\,\|\phi - u_0\|$.

Proof If $f \in (L^\infty)_1 \cap N^\perp$ then the function $K(x,y)f(y)$ is a $\lambda\times\lambda_0$-integrable function of $(x,y) \in X\times X$ and $\int fg\,d\lambda_0 = 0$. Therefore by changing the order of integration we obtain the equation

$$\int f(y)\mu(y)\,d\lambda_0(y) = \lambda(T_K f). \tag{1}$$

There is an isometric isomorphism Φ: $L^\infty \cap N^\perp \to (L^1/N)^\perp$ defined by

$$\Phi(f)(\nu+N) = \int f(y)\nu(y)\,d\lambda_0(y)$$

(for all $f \in L^\infty \cap N^\perp$ and $\nu \in L^1$). Therefore

$$\begin{aligned}\int|\mu(y)|\,d\lambda_0(y) &= \|\mu\|_1 \\ &\ge \inf\{\|\mu-h\|_1 : h \in N\} \qquad (2)\\ &= \|\mu+N\|\end{aligned}$$

$$= \sup\{\int f\mu \, d\lambda_0 : f \in (L^\infty)_1 \cap N^\perp\}$$

$$\geq \int f_0 \, \mu \, d\lambda_0 . \qquad (3)$$

There is equality at (2) if and only if condition I(c) is satisfied. There is equality at both (2) and (3) if and only if condition II(ii) is satisfied. Therefore, by (1), condition II(ii) is equivalent to condition I(c) together with the condition:

$$\lambda(T_K f) \leq \lambda(T_K f_0) \text{ for all } f \in (L^\infty)_1 \cap N^\perp . \qquad (4)$$

Conditions (4) and II(i) are together equivalent to the condition:

$$\lambda(u) \leq \lambda(u_0) \text{ for all } u \in W. \qquad (5)$$

Condition II(iii) is equivalent to the condition

$$\lambda(u_0) < \lambda(v) \text{ for all } v \in B(\phi, \|\phi - u_0\|) . \qquad (6)$$

Conditions (5) and (6) are together equivalent to the condition:

$$\lambda(u) < \lambda(v) \text{ for all } u \in W \text{ and all } v \in B(\phi, \|\phi - u_0\|),$$

which is equivalent to the two conditions I(a) and I(b). The conclusion of the theorem now follows.

Consequences of conditions II(i),(ii) and (iii). Suppose that these conditions are satisfied. Condition II(iii) is equivalent to

$$(\phi - u_0)(x) = \|\phi - u_0\| \text{ for all } x \in \text{supp}(\lambda^+), \text{ and}$$

$$(\phi - u_0)(x) = -\|\phi - u_0\| \text{ for all } x \in \text{supp}(\lambda^-),$$

where λ^+ and λ^- are the positive and negative parts of the measure λ and $\text{supp}(\lambda^+)$, $\text{supp}(\lambda^-)$ denote the supports of the measures λ^+ and λ^-. Therefore $\text{supp}(\lambda^+) \ \text{supp}(\lambda^-) = \emptyset$. Thus the function $u_0 = k_0 + T_K f_0$ is uniquely determined on $\text{supp}\lambda$, and f_0 is determined, by II(ii), λ_0-almost everywhere on $X \setminus \mu^{-1}(0)$.

If $X = [0,1]$ then the continuous function $\phi - u_0$ alternates only finitely many times between $\|\phi - u_0\|$ and $-\|\phi - u_0\|$, and consequently the measure λ 'alternates' only finitely many times. Thus there exist $\varepsilon \in \{-1,1\}$, a positive integer m and non-empty closed sets $D_1, D_2, \ldots, D_m$ such that $\sup D_i < \inf D_{i+1}$ for $i = 1,\ldots,m-1$, and

$$\text{supp}(\varepsilon\lambda)^+ = D_1 \cup D_3 \cup \ldots ,$$

$$\text{supp}(\varepsilon\lambda)^- = D_2 \cup D_4 \cup \ldots .$$

If $M = P_{r-1}$ then a simple and well-known argument shows that $m \geq r+1$ if condition II(i) is satisfied. This situation, with $K = K_r$, is the subject of Section 3.

If $M = N = \{0\}$ and K is strictly sign regular that it can be shown by a standard 'total positivity type' argument that μ has a finite number of zeros and Glashoff's conclusion follows immediately. The discussion of this section reduces in this case to that of Glashoff.

3. On the zeros of certain functions defined by K_r

We will use the notations

$$\mathcal{K}_r\begin{pmatrix} s_1,\dots\dots\dots\dots\dots,s_\rho \\ 0,\dots,r-1;t_1,\dots,t_{\rho-r}\end{pmatrix} = \begin{vmatrix} 1 & s_1\dots s_1^{r-1} & K_r(s_1,t_1)\dots K_r(s_1,t_{\rho-r}) \\ \vdots & \vdots & \vdots \\ 1 & s_\rho\dots s_\rho^{r-1} & K_r(s_\rho,t_1)\dots K(s_\rho,t_{\rho-r})\end{vmatrix}$$

and

$$\Lambda'_\rho = \{(s_1,\dots,s_\rho) : \quad 0 \leq s_1 < s_2 < \dots < s_\rho \leq 1\}.$$

If it is specified that $(t_1,\dots,t_{\rho-r}) \in \Lambda'_{\rho-r}$ then it will be convenient to interpret $t_i = -\infty$ if $i \leq 0$ and $t_i = \infty$ if $i \geq \rho-r+1$. We state a property of the determinants $\mathcal{K}_r$ which is known. It is contained in a slightly more general result which, together with proof and references can be found in SCHUMAKER.

Let $r \geq 2$, $\rho \geq r$ and $(t_1,\dots,t_{\rho-r}) \in \Lambda'_{\rho-r}$, $(s_1,\dots,s_\rho) \in \Lambda'_\rho$.

Then

$$\mathcal{K}_r\begin{pmatrix} s_1,\dots\dots\dots\dots\dots,s_\rho \\ 0\ ,\dots,r-1;t_1,\dots,t_{\rho-r}\end{pmatrix} \geq 0.$$

There is strict inequality if and only if

$$(s_1,\dots,s_\rho) \in \prod_{i=1}^{\rho} (t_{i-r},t_i).$$

We require two Propositions which are immediate consequences of this result. In both $r \geq 2$.

Proposition 1. Let $(t_1,\dots,t_{\rho+1-r}) \in \Lambda'_{\rho+1-r}$ and $(s_1,\dots,s_\rho) \in \Lambda'_\rho$ with $s_1 < t_1$ and $t_{\rho+1-r} < s_\rho$. Let

$$P(s) = \mathcal{K}_r \begin{pmatrix} s_1 \ldots\ldots\ldots\ldots s_\rho, s \\ 0, \ldots, r-1; t_1, \ldots, t_{\rho+1-r} \end{pmatrix}.$$

Then

(i) $P(s) = \sum_{i=0}^{r-1} \alpha_i s^i + \sum_{i=1}^{\rho+1-r} \beta_i K_r(s,t_i)$, where

$(-1)^{i+\rho+r-1} \beta_i \geq 0$ for $i = 1,\ldots,\rho+1-r$,

(ii) For each $i = 1,\ldots,\rho+1$, if $s_{i-1} < s < s_i$ then $(-1)^{\rho+1+i} P(s) \geq 0$,

(iii) $s_1,\ldots,s_\rho$ are the only zeros of P in $[0,1]$ if and only if $(s_1,\ldots,s_\rho) \in \prod_{i=1}^{\rho} (t_{i+1-r}, t_i)$.

<u>Proposition 2</u>. Let $(t_1,\ldots,t_\sigma) \in \Lambda'_\sigma$ and $(s_1,\ldots,s_{\sigma+r+1}) \in \Lambda'_{\sigma+r-1}$ with $s_1 < t_1$ and $t_\sigma < s_{\sigma+r+1}$. Let

$$Q(t) = \mathcal{K}_r \begin{pmatrix} s_1, \ldots\ldots\ldots\ldots s_{\sigma+r+1} \\ 0, \ldots, r-1; t_1, \ldots, t_\sigma, t \end{pmatrix}.$$

Then

(i)' $Q(t) = \sum_{i=1}^{\sigma+r+1} \gamma_i K_r(s_i,t)$, where

$(-1)^{r+\sigma+i} \gamma_i \geq 0$ for $i = 1,\ldots,\sigma+r-1$, and

$\sum_{i=1}^{\sigma+r+1} \gamma_i s_i^j = 0$ for $j = 0,\ldots,r-1$,

(ii)' For each $i = 1,\ldots,\sigma+1$, if $t_{i-1} < t < t_i$ then $(-1)^{\sigma+1+i} Q(t) \geq 0$; if $t \leq s_i$ or $s_{\sigma+r+1} \leq t$ then $Q(t) = 0$, and

(iii)' $t_1,\ldots,t_\sigma$ are the only zeros of Q in $(s_1, s_{\sigma+r+1})$ if and only if $s_i \in (t_{i-r}, t_{i-1})$ for $i = 2,\ldots,\sigma+r$.

Results concerning the existence of functions such as P and Q (of forms as in (i) and (i)') with prescribed alternation properties (as in (ii) and (ii)') are a feature of the theory of total positivity and of Chebyshev systems. A general result is given in BROWN (Lemma 2.3.4). The two Propositions differ from the general results in those clauses (iii) and (iii)' which specify precisely when the functions have only finitely many zeros.

The next theorem is related to results concerning "variation diminishing" properties of totally positive kernels (v. KARLIN). It gives precise information about the zeros and sign changes of functions of the form

$$\mu(t) = \int_0^1 K_r(s,t)d\lambda(s) ,$$

where λ is a measure orthogonal to P_{r-1} with finitely many alternations of sign on $[0,1]$.

Theorem 2. Let

$$\mu(t) = \int_0^1 K_r(s,t)d\lambda(s)$$

where $r \geq 2$ and λ is a measure on $[0,1]$ such that

(i) $\lambda(k) = 0$ for all $k \in P_{r-1}$, and

(ii) for some $m \geq r+1$, $\varepsilon \in \{-1,1\}$ and non-empty closed sets $D(1),\ldots,D(m)$,

$$\sup D(i) < \inf D(i+1) \text{ for } i = 1,m-1,$$
$$\operatorname{supp}(\varepsilon\lambda)^+ = D(1) \cup D(3) \cup \ldots,$$
$$\operatorname{supp}(\varepsilon\lambda)^- = D(2) \cup D(4) \cup \ldots.$$

Then there exist $j_0 \in \{0,\ldots,m-r-1\}$ and $n \in \{r+1,\ldots,m-j_0\}$ such that on the interval $(\sup D(j_0+1), \inf D(j_0+n))$ the function μ has exactly $n-r-1$ sign changes, has at most $(n-r-1) + \max\{0,m-r-2\}$ zeros, and has the sign of $(-1)^{r+j_0}\varepsilon$ immediately to the right of $\sup D(j_0+1)$.

The proof of the theorem will be achieved by two lemmas. The first is proved by an argument (not original to this author) which can be used to obtain most (if not all) of those results in the theory of total positivity which concern "variation diminishing" properties of kernels. The assumptions of Lemma 1 are those of Theorem 2.

Lemma 1. If

$$(t_1,\ldots,t_{m-r}) \in \Lambda'_{m-r} \cap \prod_{i=1}^{m-r} (\sup D(i), \inf D(i+r)) \tag{7}$$

then

$$(-1)^{i+r}\varepsilon\mu(t_i) < 0 \text{ for at least one } i \in \{1,\ldots,m-r\} \tag{8}$$

Proof Suppose that (7) holds but that (8) is false. Then for $i = 1,\ldots,m-1$, we can choose s_i in the set $(\sup D(i),t_i) \cap (t_{i+1-r}, \inf D(i+1))$ which is non-empty. Then

$$(s_1,\ldots,s_{m-1}) \in \Lambda'_{m-1} \cap \prod_{i=1}^{m-1} (t_{i+1-r},t_i) \cap \prod_{i=1}^{m-1} (\sup D(i), \inf D(i+1)).$$

Let

$$P(s) = \mathcal{K}_r \begin{pmatrix} s_1, \dots\dots\dots\dots, s_{m-1}, s \\ 0, \dots, r-1; t_1, \dots, t_{m-r} \end{pmatrix}$$

$$= \sum_{i=0}^{r-1} \alpha_i s^i + \sum_{i=1}^{m-r} \beta_i K_r(s, t_i).$$

All the conditions and conclusions of Proposition 1 ($\rho = m-1$) are satisfied and therefore

$$0 \geq \sum_{i=1}^{m-r} (-1)^{i+m-r} \beta_i . (-1)^{i+r+1} \varepsilon\, \mu(t_i)$$

$$= (-1)^{m+1} \varepsilon \sum_{i=1}^{m-r} \beta_i \int_0^1 K_r(s, t_i) d\lambda(s)$$

$$= \int (-1)^{m+1} P(s) . \varepsilon d\lambda(s)$$

$$> 0 ,$$

which is a contradiction. This proves the lemma.

It now has to be shown that the conclusion of Theorem 2 follows from Lemma 1. The proof is by induction on the integer m. The proof is by induction on the integer m. The second lemma has to be stated in a form appropriate to the proof. The theorem will follow from the lemma by reading $\bar{\xi}(i) = \sup D(i)$ and $\underline{\xi}(i) = \inf D_i$.

Lemma 2. Suppose than an integer m, points

$$0 \leq \bar{\xi}(1) \leq \underline{\xi}(2) \leq \bar{\xi}(2) \leq \dots \leq \bar{\xi}(m-1) \leq \underline{\xi}(m) \leq 1,$$

a continuous function μ and $\varepsilon \in \{-1,1\}$ have the properties

(i) $m \geq r+1$,

(ii) $\bar{\xi}(i) < \underline{\xi}(i+r)$ for $i = 1, \dots, m-r$, and

(iii) if $(t_1, \dots, t_{m-r}) \in \Lambda'_{m-r} \cap (\bar{\xi}(i), \underline{\xi}(i+r))$ then $(-1)^{i+r} \varepsilon \mu(t_i) < 0$ for at least one $i \in \{1, \dots, m-r\}$.

Then there exist $n \in \{r+1, \dots, m\}$ and $j_0 \in \{0, \dots, m-n\}$ such that on the interval $(\bar{\xi}(j_0+1), \underline{\xi}(j_0+n))$ the function μ has exactly $n-r-1$ sign changes, at most $(n-r-1) + \max\{0, m-r-2\}$ zeros and the sign of $(-1)^{r+j_0}\varepsilon$ immediately to the right of $\bar{\xi}(j_0+1)$.

Proof is by induction on m.

The case $m = r+1$. Property (iii) asserts that $(-1)^r \varepsilon\, \mu(t) > 0$ for all $t \in (\bar{\xi}(1), \underline{\xi}(1+r))$. Thus the conclusion of the lemma holds with $j_0 = 0$ and $n=m=r+1$.

The inductive step Suppose that $m > r+1$ and that the statement of the lemma is true when m is replaced by $m-1$.

Suppose that the hypotheses of the lemma are satisifed. If $(-1)^r \varepsilon\mu(t) > 0$ for all $t \in (\bar{\xi}(1), \underline{\xi}(r+1))$ then the conclusion is satisfied by $j_0 = 0$ and $n = r+1$. Therefore we may suppose that $(-1)^r \varepsilon\mu(t) \le 0$ for some $t \in (\bar{\xi}(1), \underline{\xi}(r+1))$. Let

$$x = \inf\{t \in (\bar{\xi}(1), \underline{\xi}(r+1)) : (-1)^r \varepsilon\mu(t) \le 0\}.$$

Then either $x \in (\bar{\xi}(1), \underline{\xi}(r+1))$, $\mu(x) = 0$ and $(-1)^r \varepsilon\mu(t) > 0$ for all $t \in (\bar{\xi}(1), x)$ or $x = \bar{\xi}(1)$.

If now $x < t_2 < \ldots < t_{m-r}$ and $\bar{\xi}(i) < t_i < \underline{\xi}(i+r)$ for $i = 2, \ldots, m-r$ then we can choose t_1, either $t_1 = x$ if $x > \bar{\xi}(1)$ or $t_1 \in (\bar{\xi}(1), t_2)$ if $x = \bar{\xi}(1)$, such that $t_1 \in (\bar{\xi}(1), t_2)$ and $(-1)^{r+1} \varepsilon\mu(t_1) \ge 0$. Therefore, but property (iii), $(-1)^{i+r} \varepsilon\mu(t_i) < 0$ for at least one $i \in \{2, \ldots, m-r\}$.

Now let $\bar{\xi}'(j) = \max\{\bar{\xi}(j+1), x\}$ and $\underline{\xi}'(j+1) = \max\{\underline{\xi}(j+2), x\}$ for $j = 1, \ldots, m-2$. Then the hypotheses of the lemma are satisfied by m-1 (in place of m), the points

$$0 \le \bar{\xi}'(1) \le \underline{\xi}'(2) \le \bar{\xi}'(2) \le \ldots \le \bar{\xi}'(m-2) \le \underline{\xi}'(m-1) \le 1,$$

the function μ and $-\varepsilon$ (in place of ε). Now, by the inductive assumption there exist $n' \in \{r+1, \ldots, m-1\}$ and $j_0' \in \{0, \ldots, m-1-n'\}$ such that on the interval $(\bar{\xi}'(j_0'+1), \underline{\xi}'(j_0'+n'))$ the function μ has exactly $n'-r-1$ sign changes, at most $(n'-r-1) + \max\{0, m-r-3\}$ zeros and the sign of $(-1)^{r+j_0'+1}\varepsilon$ immediately to the right of $\bar{\xi}'(j_0'+1)$. The conclusion of the lemma now follows. If $x \le \bar{\xi}(j_0'+2)$ then the conclusion holds with $j_0 = j_0' + 1$ and $n = n'$. If $\bar{\xi}(j_0'+2) < x$ and j_0' is even then μchanges sign at x and the conclusion of the lemma is satisfied by $j_0 = j_0'$ and $n = n'+1$. If $\bar{\xi}(j_0'+2) < x$ and j_0' is odd then $\mu(x) = 0$ but μ does not change sign at x; the conclusion of the lemma is satisfied by $j_0 = j_0' + 1$ and $n = n'$.

The proofs of Lemma 2 and Theorem 2 are complete.

4. Sattes' Theorem

It is now possible to give a proof of Sattes' theorem concerning best uniform approximation by smooth functions which is very different from the proof given by Sattes. The theorem asserts that if u_0 is a best approximation to $\phi \in C([0,1])$ from $W_{\infty,r}$, where $r \ge 2$, then there is an interval $[\alpha, \beta]$ such that on $[\alpha,\beta]$ u_0 is a uniquely determined perfect spline. Furthermore the best approximations u_0 are characterised by the behaviour of ϕ and u_0 on an interval $[\alpha,\beta]$. Sattes' theorem will be stated in a very slightly modified form.

Theorem 3 (SATTES) Let $r \geq 2$. Suppose $\phi \in C[0,1]\backslash W_{\infty,r}$ and $u_0 \in W_{\infty,r}$. Consider the elements

$$\left.\begin{aligned} &[\alpha,\beta] \subseteq [0,1], \quad n \geq r+1, \quad \varepsilon \in \{-1,1\}, \\ &\alpha = \xi_1 < \xi_2 < \ldots < \xi_n = \beta, \\ &\alpha = \tau_0 < \tau_1 < \ldots < \tau_{n-r-1} < \tau_{n-r} = \beta \end{aligned}\right\} \tag{9}$$

and the conditions

(i) $(\phi - u_0)(\xi_i) = (-1)^i \varepsilon \, \|\phi - u_0\|$ for $i = 1,\ldots,n$;

(ii) $u_0^{(r)}(t) = (-1)^{r+i}\varepsilon$ for a.e. $t \in (\tau_{i-1},\tau_i)$ and $i = 1,\ldots,n-r$;

(iii) $\xi_{i+1} < \tau_i < \xi_{i+r}$ for $i = 1,\ldots,n-r-1$.

(a) If u_0 is a best approximation to ϕ from $W_{\infty,r}$ then there exist elements as in (9) such that conditions (i) and (ii) are satisfied.

(b) If there exist elements as in (9) such that (i) and (ii) are satisfied then there exist elements as in (9) such that (i),(ii) and (iii) are satisfied.

(c) If elements as in (9) are such that conditions (i),(ii) and (iii) are satisfied then u_0 is a best approximation to ϕ from $W_{\infty,r}$ and $u_0'|[\alpha,\beta] = u_0|[\alpha,\beta]$ for every best approximation u_0' to ϕ from $W_{\infty,r}$.

Proof The function u_0 is of the form

$$u_0(s) = k_0(s) + \int_0^1 K_r(s,t) f_0(t)\,dt$$

where $k_0 \in P_{r-1}$ and $f_0 = u_0^{(r)} \in (L^\infty([0,1]))_1$.

Suppose that u_0 is a best approximation to ϕ from $W_{\infty,r}$. We can choose $\lambda \in C[0,1]^* \backslash\{0\}$ to satisfy condition I(b) of Theorem 1. The subspace $N = \{0\}$, so the function

$$\mu(t) = \int_0^1 K_r(s,t)\,d\lambda(s)$$

satisfies condition I(c). Therefore conditions II of Theorem 1 are satisfied. By the remarks following the proof of Theorem 1 the function μ satisfies the hypotheses of Theorem 2. Assertion (a) of Theorem 3 now follows from Theorem 2: $\alpha = \sup D(j_0+1)$, $\beta = \inf D(j_0+n)$, the point ξ_i is chosen from $D(j_0+i)$ for $i = 2,\dots,n-1$, and the points $\tau_1,\dots,\tau_{n-r-1}$ are the points of (α,β) at which μ changes sign.

The proof of the assertion (b) is trivial. Suppose that elements as in (9) satisfy conditions (i) and (ii). Let k be the largest of $0,\dots,n-r-1$ such that $\tau_k \le \xi_{k+1}$. Then $\xi_{k+i+1} < \tau_{k+i}$ for $i = 1,\dots,n-k-r-1$. Let ℓ be the smallest of $1,\dots,n-k-r$ such that $\xi_{k+\ell+r} \le \tau_{k+\ell}$. Then $\tau_{k+i} < \xi_{k+i+r}$ for $i = 1,\dots,\ell-1$. It now follows that conditions (i),(ii) and (iii) are satisfied by $[\alpha',\beta'] = [\xi_{k+1},\xi_{k+\ell+r}]$ in place of $[\alpha,\beta]$, $\ell+r$ in place of n, and the points $\alpha' = \xi_{k+1} < \xi_{k+2} < \dots < \xi_{k+\ell+r} = \beta'$ and $\alpha' < \tau_{k+1} < \dots < \tau_{k+\ell-1} < \beta'$.

It remains to prove assertion (c). Suppose that conditions (i),(ii) and (iii) are satisfied by elements as in (9). Let

$$\mu(t) = (-1)^n \varepsilon \mathcal{K}_r \begin{pmatrix} \xi_1,\dots\dots\dots\dots,\xi_n \\ 0,\dots,r-1;\tau_1,\dots,\tau_{n-r-1},t \end{pmatrix}$$

$$= (-1)^n \varepsilon \sum_{i=1}^{n} \gamma_i K_r(\xi_i,t).$$

The properties of μ and the coefficients γ_i $(i = 1,\dots,n)$ which we require are those given by (i)',(ii)' and (iii)' of Proposition 2 ($Q = \mu$, $s_i = \xi_i$, $t_i = \tau_i$). Let a measure $\lambda \in C^*([0,1]) \setminus \{0\}$ be defined by

$$\lambda = \sum_{i=1}^{n} (-1)^n \varepsilon \gamma_i \delta_{\xi_i},$$

(where δ_ξ is unit measure at ξ). It now follows that ϕ, u_0, λ, μ satisfy the conditions II of Theorem 2 applied to this situation. (II(i) is a consequence of (i)'. II(ii) is a consequence of (ii) and (ii)', II(iii) is a consequence of (i) and (i)'. It now follows from Theorem 1 that conditions I(a) and I(b) are satisfied (I(c) is vacuous). So u_0 is a best approximation to ϕ from $W_{\infty,r}$. If u_0' is also a best approximation then it also satisfies II(ii) and II(iii). Consequently $D^r(u_0-u_0')$ is almost everywhere zero on $[\alpha,\beta]$ and $(u_0-u_0')|[\alpha,\beta]$ is a polynomial of degree $\le r-1$. However $(u_0-u_0')(\xi_i) = 0$ for $i = 1,\dots,n$ and $n \ge r+1$. Therefore $(u_0 - u_0')|[\alpha,\beta] = 0$. This completes the proof of Sattes' theorem.

Finally we state a corollary and a problem.

<u>Corollary</u> Suppose $\phi \in C([0,1]) \setminus W_{\infty,r}$. Then there exists a measure $\lambda_0 \in C([0,1])^*$

which is a convex combination of extreme points of the unit ball of $C([0,1])^*$ and which separates $B(\phi,d(\phi,W_{\infty,r}))$ and $W_{\infty,r}$.

Proof If λ is the measure constructed in the proof of (c) of Theorem 3 then $\lambda_0 = \|\lambda\|^{-1}\lambda$ is such a measure.

The proof of the corollary depends on the entire paper. One might hope to attain a more direct proof and perhaps thereby greater insight into the general problem.

5. References

Brown, A.L. (1982) Finite rank approximations to integral operators which satisfy certain total positivity conditions. J. Approximation Theory 34, 42-90.

Glashoff, K. (1980) Restricted approximation by strongly sign-regular kernels: The finite bang-bang principle. J. Approximation Theory 29, 212-217.

Karafiat, A. (1977) The problem of the number of switches in parabolic equations with control. Ann. Polon. Math. 34, 289-316.

Karlin, S. (1968) Total Positivity, Vol. 1. (Stanford Univ. Press, Stanford).

Pinkus, A. (1980) Best approximation by smooth functions. Technion Preprint series No. MT-487.

Sattes U. (1980) Best Chebyshev approximation by smooth functions. Quantitative Approximation (edited by R.A. Devore and K. Scherer) Proc. Int. Symp., Bonn 1979, 279-289. (Academic Press, New York).

Schumaker, L.L. (1981) Spline functions: Basic Theory (John Wiley, New York).

A.L. Brown, School of Mathematics, The University of Newcastle upon Tyne, Newcastle upon Tyne NE1 7RU, U.K.

International Series of
Numerical Mathematics, Vol. 72

ON PARAMETRIC INFINITE OPTIMIZATION

A. R. da Silva

Instituto de Matemática, Universidade Federal de Rio de Janeiro, Rio de Janeiro, Brazil

Abstract

We consider infinite parametric problems and show that the best approximation problems in real normed spaces and the parametric semi-infinite problems can be reformulated as parametric infinite problems. Further we give sufficient conditions for the lower semicontinuity of the optimal set mapping in the linear case.

1. Introduction and Problem Formulation

Let Y be a Hausdorff topological space, U be a non-empty subset of Y, X be a partially ordered real normed space, and K the convex cone which gives the partial order of X. Let $p : U \to \mathbb{R}$ and $C : U \to X$ be continuous mappings. For each parameter $\sigma := (C,p)$, we consider the following optimization problem:

$$\underline{\text{MPI}(\sigma)} \qquad \text{Minimize } \; p : U \to \mathbb{R}$$
$$\text{subject to } \; C(u) \leq 0, \quad u \in U$$

In this way, we get a family of optimization problems each of them depending on the choice of the parameter σ. For each parameter σ, we define

(i) the set of feasible points

$$Z_\sigma := \{u \in U : C(u) \leq 0\} ;$$

(ii) the minimal value

$$E_\sigma := \inf\{p(u) \in \mathbb{R} : u \in Z_\sigma\} ;$$

(iii) the set of minimal points

$$P_\sigma := \{u \in U : p(u) = E_\sigma\} .$$

In section 2, we reformulate the problem MPI(σ) by imposing certain conditions on the cone K and its dual cone.

In section 3, we show that semi-infinite problems and best approximation problems in real normed spaces are of type MPI(σ). Hence with this approach it is possible not only to derive new results in the parametric infinite optimization theory, but also give an unified treatment of certain problems in the parametric optimization and approximation theory.

In section 4, we prove an always sufficient criterion for the lower semicontinuity of the mapping P_σ, in the linear case.

We denote by X^* the topological dual of a given real normed linear space X and by K^* the dual cone of a given convex cone K, i.e., $K^* := \{\varphi \in X^*;\ \varphi(k) \geq 0,$ for all $k \in K\}$. All concepts not explicitly defined can be found in BROSOWSKI [2] or in HOLMES [4].

2. Reformulation of the Problem

We recall that if K is a convex cone in a real normed linear space, which has a non-empty interior then its dual cone K^* is always a w*-closed convex cone in X^*.

Lemma 2.1 (KLEE). Let K^* be a dual cone and τ be a locally convex topology on a dual space X^*. Then K^* is a τ-locally compact cone if and only if K^* has a τ-compact base.

For a proof see e.g. HOLMES [4].

Lemma 2.2. Let $K \subset X$ be a closed convex proper cone such that K^* is a w*-locally compact cone. Then the following are equivalent:

(i) $x \in -K$;

(ii) $k^*(x) \leq 0$ for all $k^* \in K^*$;

(iii) $e^*(x) \leq 0$ for all $e^* \in \text{ext}(B^*)$, where B^* is a w*-compact base for K^*.

Proof. The implications (i) $\Rightarrow$ (ii) $\Rightarrow$ (iii) are trivial. To prove (iii) $\Rightarrow$ (i) take $x \in X$ with $e^*(x) \leq 0$ for all $e^* \in \text{ext}(B^*)$.

By the Krein-Milman's theorem $B^* = \overline{\text{conv}}^{w^*}(\text{ext } B^*)$. Hence

(*) $b^*(x) \leq 0$ for every $b^* \in B^*$.

Now if $x \notin -K$, there exist $\varphi \in X^*$ and $\alpha \in \mathbb{R}$ such that $\varphi(p) \leq \alpha < \varphi(x)$ for every $p \in -K$. In particular $\varphi(x) > 0$.

φ is lower bounded therefore $\varphi \in K^*$. Since B^* is a base for K^*, there exist a unique $b_1^* \in B^*$ and $\lambda > 0$ such that $\varphi = \lambda b_1^*$ whence $b_1^*(x) > 0$. But this contradicts (*).

<u>Theorem 2.3</u>. Consider the minimization problem $MPI(\sigma)$. If K is a closed convex proper cone, such that K^* is a w*-locally compact cone, then $MPI(\sigma)$ and the minimization problem

<u>$MPI_e(\sigma)$</u> Minimize $p : U \to \mathbb{R}$

subject to $e^*(C(u)) \leq 0$ for every $e^* \in ext(B^*)$

are equivalent. That is

(i) $Z_\sigma = Z_\sigma^e := \{u \in U : e^*(C(u)) \leq 0$ for every $e^* \in ext(B^*)\}$,

(ii) $P_\sigma = P_\sigma^e := \{u \in Z_\sigma^e \mid p(u) = E_\sigma\}$

and

(iii) $Z_\sigma^< := \{u \in U \mid C(u) \in -(\text{int } K)\}$ is identical with

$Z_\sigma^{e<} := \{u \in U \mid e^*(C(u)) < 0$ for every $e^* \in ext(B^*)\}$.

<u>Proof</u>. (i) and (ii) follow immediately from lemma 2.2. To prove (iii) let $u \in Z_\sigma^<$ and assume there exists $e_0^* \in ext(B^*)$ such that $e_0^*(C(u)) = 0$. In this case, since $C(u) \in -int(K)$, for each $k \in K$ we can find $\lambda > 0$ such that $C(u) + \lambda k \in -K$.

Hence $e_0^*(k) = 0$, for each $k \in K$. Now since K^* is w*-locally compact, int $K \neq \phi$; whence $X = K - K$, thus $e_0^* \equiv 0$ which is impossible. Therefore

$$e^*(C(u)) < 0 \quad \text{for every} \quad e^* \in ext(B^*) .$$

To prove the converse take $u \in Z_\sigma^{e<}$ and assume $C(u) \notin -(\text{int } K)$. In this case there exist $\varphi \in X^*$ and $\alpha \in \mathbb{R}$ such that

$$\varphi(-k) \leq \alpha \leq \varphi(C(u)) \quad \text{for every} \quad k \in K .$$

Now $\varphi \in K^*$, hence there exist unique $b^* \in B^*$ and $\rho > 0$ such that $\varphi = \rho b^*$. Thus $b^*(C(u)) \geq 0$.

On the other hand, since $u \in Z_\sigma^{e<}$, $b^*(C(u)) \leq 0$. Thus $b^*(C(u)) = 0$.

Now let $Z^* := \{\Psi \in B^* \mid \Psi(C(u)) = 0\}$. The convex set Z^* is a non-empty w*-compact extremal subset of B^*, hence $\operatorname{ext} Z^* \neq \phi$ and $\operatorname{ext}(Z^*) = Z^* \cap \operatorname{ext}(B^*)$ which is a contradiction. That proves our claim.

3. <u>Some examples of problems of type MPI(σ)</u>

3.1. Best approximation in normed spaces

Let Z be a real normed space, V be a non-empty subset of Z, z be a point of $Z \setminus \bar{V}$ and U be defined as $U := V \times \mathbb{R}$; further let us define $X := Z \times \mathbb{R}$ and $K := \{(w,r) \in X : r \geq \|w\|\}$. If we define $|||(w,r)||| := \|w\| + |r|$ and

$$(w,r) \in K \quad \text{iff} \quad r \geq \|w\| ,$$

then (X,K) becomes a partially ordered real normed space. If $p : U \to \mathbb{R}$ and $C_z : U \to X$ are defined by $p(v,r) := r$ and $C_z(v,r) := (-v,-r) + (z,0)$ then p and C_z are continuous mapping and the problem

$$\text{BA}(z) \qquad \text{Minimize } p : U \to \mathbb{R}$$
$$\text{subject to } \|v - z\| \leq r$$

can be written as

$$\text{MPI}(\sigma_z) \qquad \text{Minimize } p : U \to \mathbb{R}$$
$$\text{subject to } C_z(u) \leq 0 .$$

Now an easy calculation shows that $\bar{v} \in P_V(z)$ (where $P_V(z) := \{\tilde{v} \in V : \|\tilde{v} - z\| \leq \|v - z\| \ \forall\, v \in V\}$) if and only if the problem BA(z) has the point $(\bar{v}, \|\bar{v} - z\|)$ as a solution.

Further, it is easy to show that K^* is w*-locally compact.

In fact, $B_{Z^*} \times \{1\}$ is a base for K^*. Here B_{Z^*} denotes the unit ball of Z^*. Hence we can rewrite BA(z) (and therefore MPI(σ_z)) as

$$\text{BA}_e(z) \qquad \text{Minimize } p : U \to \mathbb{R}$$
$$\text{subject to } e^*(z) - r \leq e^*(v), \text{ where } e^* \in \operatorname{ext}(B_{Z^*}) .$$

The next example shows that semi-infinite problems are of type $MPI(\sigma)$.

3.2. Semi-infinite Optimization

Let now T be a compact Hausdorff space, $X = C(T)$ ordered by the closed cone $K := \{x \in C(T) : x(t) \geq 0, \quad t \in T\}$. Further, let Y be a Hausdorff topological space, U a non-empty subset of Y, and $C : U \to X$ a continuous mapping given by $C(u) := A(\cdot,u) - b(\cdot)$. As it is well known, the set of all positive measures with norm equal one form a w*-compact base B^* for the dual cone K^*.

Further the extremal points of this base are given by the positive point evaluation functionals ε_t, $t \in T$ (see e.g. HOLMES [4]).

Hence the equivalence between $MPI(\sigma)$ and $MPI_e(\sigma)$ in this case means that the side condition $C(u) \leq 0$, $u \in U$, can be rewritten as $\varepsilon_t(A(\cdot,u) - b(\cdot)) = A(t,u) - b(t) \leq 0$, where $t \in T$ and $u \in U$. Thus all semi-infinite problems are of type $MPI(\sigma)$ in which case Y is a finite-dimensional euclidean space, see e.g. BROSOWSKI [2].

4. The lower semicontinuity of the optimal set mapping in the linear case, for right-hand side parameters

Consider a real normed space Y and X as in section 1. Let p be a continuous linear functional on Y, $A : Y \to X$ be a continuous linear operator and $C : Y \to X$ be defined by $C(u) := A(u) - b$.

For each $b \in X$, we consider the problem

$$\underline{MLPI(b)} \quad \text{Minimize } p : Y \to \mathbb{R}$$
$$\text{subject to } A(u) \leq b, \quad u \in Y.$$

As shown in section 2, if K is closed and K^* is w*-locally compact, we can rewrite $MLPI(b)$ as

$$\underline{MLPI_e(b)} \quad \text{Minimize } p : Y \to \mathbb{R}$$
$$\text{subject to } e^*(A(u)) \leq e^*(b) \text{ for every } e^* \in \text{ext}(B^*).$$

Define $L_{A,p} := \{b \in X : P_b \neq \phi\}$, where we simply write P_b instead of P_σ.

Definition 4.1. Let $b \in L_{A,p}$. For each $v \in P_b$, define

$$\Sigma_{b,v} := \{b^* \in B^* : b^*(A(v)) = b^*(b)\} ,$$

$$E_{b,v} := \{e^* \in \text{ext}(B^*) : e^*(A(v)) = e^*(b)\} ,$$

and

$$N_b := \bigcap_{u \in P_b} \{e^* \in \text{ext}(B^*) : e^*(A(u)) = e^*(A(v))\} .$$

The following lemmata can be easily proved.

<u>Lemma 4.2.</u> $\Sigma_{b,v}$ is a non-empty extremal subset of B^*.

<u>Lemma 4.3.</u> $E_{b,v} = \text{ext}(\Sigma_{b,v}) = \Sigma_{b,v} \cap \text{ext}(B^*)$.

<u>Lemma 4.4.</u> If $\dim P_b < \infty$ then

$$E_{b,v} = \bigcap_{u \in P_b} E_{b,u} \quad \text{for every } v \in \text{rel int}(P_b) .$$

<u>Theorem 4.5.</u> For each $b \in L_{A,p} \subset X$ suppose that

(i) $\dim P_b < \infty$;

(ii) there exists a neighborhood $U_b \subset L_{A,p}$ and a mapping $\pi : U_b \to U$ such that

(a) $\pi(b') \in P_b$ for every $b' \in U_b$,

(b) If the sequence (b_n) in $L_{A,b}$ converges to b, then the set $\{\pi(b_n) \in U : n \in \mathbb{N}\}$ is relatively compact, i.e. π is a selection on U_b which transforms convergent sequences to relatively compact sets.

(iii) P is closed at b;

(iv) there exists a w*-open convex set $W^* \subset X^*$ such that

$$N_b \supset W^* \cap \text{ext}(B^*) \supset \bigcap_{u \in P_b} E_{b,u} .$$

Then the optimal set mapping P is lower semicontinuous.

Proof. Assume P is not lower semicontinuous at $\bar{b} \in L_{A,p}$. Then there exist a sequence (b_n) in $L_{A,p}$, an element $v \in P_{\bar{b}}$, and an open neighborhood V of v such that

$$b_n \to \bar{b} \quad \text{and} \quad P_{b_n} \cap V = \phi \quad \text{for every} \quad n \in \mathbb{N} .$$

Since $\dim P_{\bar{b}} < \infty$ we can assume $v \in \text{rel int}(P_{\bar{b}})$. By (ii) we can choose for each $n \in \mathbb{N}$ an element $\pi(b_n)$ in P_{b_n} such that the set $G := \{\pi(b_n) \in U \mid n \in \mathbb{N}\}$ is relatively compact. Then G has an accumulation point w in U, which is in $P_{\bar{b}}$ by condition (iii). Since $P_{b_n} \cap V = \phi$ we have $w \neq v$.

Without loss of generality we can assume that $\pi(b_n)$ converges to w. Now the sequence $v_n := v + \pi(b_n) - w$ converges to v.

Further P_{b_n} does not contain v_n and $p(v_n) = p(\pi(b_n))$ for $n \geq n_1$ with n_1 suitably chosen.

Hence $v_n \notin Z_{b_n}$ and therefore

$$\max_{e^* \in \text{ext}(B^*)} e^*(A(v_n) - b_n) > 0 \quad \text{for every} \quad n \geq n_1 .$$

Define $H_n := A(v_n) - b_n$ and

$$R_n := \{e^* \in \text{ext}(B^*) : e^*(H_n) = \max_{f^* \in \text{ext}(B^*)} f^*(H_n)\} .$$

For each $e^* \in R_n$,

$$0 < e^*(H_n) = e^*(A(v - w)) + e^*(A(\pi(b_n)) - b_n)$$

$$\leq e^*(A(v - w)) .$$

Hence $N_{\bar{b}} \cap R_n = \phi$ for every $n \geq n_1$.

From (iv)

$$(*) \quad (W^* \cap \text{ext}(B^*)) \cap R_n = \phi \quad \text{for every} \quad n \geq n_1 .$$

On the other hand, the set $W^* \cap \text{ext}(B^*)$ is $w^*|_{\text{ext}(B^*)}$-open, where $w^*|_{\text{ext}(B^*)}$ is the relative topology induced on $\text{ext}(B^*)$ by the w^*-topology. Further

$$\bigcap_{u \in P_{\bar{b}}} E_{\bar{b},u} \subset W_1^* := W^* \cap \text{ext}(B^*) .$$

We claim that

$$\sup_{e^* \in \text{ext}(B^*) \setminus W_1^*} e^*(A(v) - \bar{b}) < 0 .$$

In fact, it is obvious that

$$\sup_{e^* \in \text{ext}(B^*) \setminus W_1^*} e^*(A(v) - \bar{b}) = \sup_{e^* \in \text{ext}(B^*) \setminus W^*} e^*(A(v) - \bar{b})$$

$$\leq \sup_{e^* \in B^* \setminus W^*} e^*(A(v) - \bar{b}) .$$

On the other hand, by Krein-Milman's theorem and lemmata 4.3 and 4.4., we have

$$\overline{\text{conv}}^{w^*} \Big(\bigcap_{u \in P_{\bar{b}}} E_{\bar{b},u} \Big) = \Sigma_{\bar{b},v} .$$

Further, by (iv) and lemma 4.4, $W^* \supset E_{\bar{b},v}$. Since W^* is convex, $W^* \supset \text{conv}\, E_{\bar{b},v}$. Moreover $W^* \supset \Sigma_{\bar{b},v}$. In fact, otherwise let $c^* \in \Sigma_{\bar{b},v} \setminus W^*$. Now since W^* is convex and w^*-open, there exists a supporting hyperplane containing c^*; hence there exists g in X such that

$$c^*(g) = \inf_{\Psi \in W^*} \Psi(g) .$$

Thus if $H := \{\Psi \in X^* : \Psi(g) = c^*(g)\}$, then

$$(**) \qquad W^* \cap H = \phi .$$

Since $\Sigma_{\bar{b},v}$ is contained in the w*-closure of W^*,

$c^*(g) = \inf_{\Psi \in \Sigma_{\bar{b},v}} \Psi(g)$, whence $H \cap \Sigma_{\bar{b},v}$ is an extremal subset of $\Sigma_{\bar{b},v}$ in H. By Lemma 4.3 $H \cap \Sigma_{\bar{b},v} \subset E_{\bar{b},v}$. But $E_{\bar{b},v} \subset W^*$, whence $H \cap W^* \neq \phi$, which contradicts (**). Consequently $W^* \supset \Sigma_{\bar{b},v}$.

Since $B^* \setminus W^*$ is w*-compact and $W^* \supset \Sigma_{\bar{b},v}$

$$\sup_{e^* \in B^* \setminus W^*} e^*(A(v) - \bar{b}) < 0 ,$$

whence

$$\sup_{e^* \in \text{ext}(B^*) \setminus W^*} e^*(A(v) - \bar{b}) < 0 .$$

Hence there exists $\varepsilon > 0$ such that $e^*(A(v) - \bar{b}) \leq -2\varepsilon$, for all $e^* \in \text{ext}(B^*) \setminus W^*$.

On the other hand

$$|b^*(A(v_n - v)) - b^*(b_n) + b^*(\bar{b})| \leq$$

$$\leq \|b^*\|(\|A\|\|v_n - v\| + \|b_n - \bar{b}\|), \quad \text{for each } b^* \in B^* .$$

Since A is continuous, we have, for every $b^* \in B^*$,

$$b^*(A(v_n) - b_n) \to b^*(A(v) - \bar{b}) \quad \text{when } n \to \infty ,$$

which implies

$$e^*(H_n) = e^*(A(v_n) - b_n) \leq -\varepsilon \quad \text{for all } e^* \in \text{ext}(B^*) .$$

Thus, we have

$$\mathrm{ext}(B^*) \setminus W^* \subset \mathrm{ext}(B^*) \setminus R_n ,$$

for suitable n or equivalently, $W^* \supset R_n$. Whence $W^* \cap \mathrm{ext}(B^*) \supset R_n$, contradicting (*). This proves the theorem.

Lemma 4.6. Consider the linear minimization problem MPLI(b). If the Slater-condition is fulfilled, i.e., there exists $u \in U$ such that $b - A(u) \in \mathrm{int}\, K$, then the optimal set mapping P is closed at b.

Proof. By assumption there exists $v \in U$ such that $b - A(v) \in \mathrm{int}\, K$ hence using (iii) in theorem 2.3

$$e^*(A(v) - b) < 0 \quad \text{for every} \quad e^* \in \mathrm{ext}(B^*) .$$

Let (b_n) be a sequence in $L_{A,p}$ such that $b_n \to b$, $u_n \in P_{b_n}$ and $w := \lim_n u_n$.

Since, for every $e^* \in \mathrm{ext}(B^*)$,

$$e^*(A(w)) = e^*(A(\lim_n u_n)) \leq \lim_n e^*(b_n) = e^*(b) ,$$

it follows that $w \in Z_b$.

This implies P closed at b, if $Z_b = P_b$. So we may assume $P_b \neq Z_b$.

Choose an arbitrary element $u \in P_b$ and define

$$v_m := (1 - 1/m)u + (1/m)v, \quad m \in \mathbb{N} .$$

We claim that for each m there exists $n_0(m) \in \mathbb{N}$ such that

$$e^*(A(v_m)) < e^*(b_n)$$

for every $n \geq n_0(m)$ and $e^* \in \mathrm{ext}(B^*)$.

In fact,

$$\begin{aligned} e^*(A(v_m)) &= (1 - 1/m)e^*(A(u)) + (1/m)e^*(A(v)) \\ &< (1 - 1/m)e^*(A(u)) + (1/m)e^*(b) \\ &< e^*(b). \end{aligned}$$

Since $b_n \to b$ and e^* is continuous, we have for every $e^* \in \text{ext}(B^*)$

$$e^*(A(v_m)) < e^*(b_n)$$

whenever $n \geq n_0(m)$.

Consider the subsequence $(u_{n_0(m)})$ of (u_n). Then we have $p(u_{n_0(m)}) \leq p(v_m) = (1 - 1/m)p(u) + (1/m)p(v)$. Hence if we let $m \to \infty$ we obtain $E_b = p(u) \geq p(w)$, which shows that $w \in P_b$ and therefore that the optimal set mapping P is closed at b.

Corollary 4.7. In the theorem 4.5 the condition (iii) can be substituted by the Slater-condition.

Remark 4.8. The theorem 4.5 contains as a special case a result of BROSOWSKI [2], who considered the case of a finite dimensional euclidean space Y and the space $X = C(T)$, with its usual order (see 3.2).

Lemma 4.9. Let Z be a real normed linear space, V a linear subspace of Z. If for each $z \in Z$ $\dim P_V(z) < \infty$, then P_V is upper semicontinuous if and only if the mapping $z \mapsto (P_V(z), d(z,V))$ is upper semicontinuous, where $d(z,V) := \inf_{v \in V} \|z - v\|$.

Proof. Necessity. Fix $z \in Z$ and let $n := \dim P_V(z)$. Let U be an open set in $\mathbb{R}^n \times \mathbb{R}$. Without loss of generality we can assume that $U = \prod_{i=1}^{n} (a_i, b_i) \times (\alpha, \beta)$. Now assume

$$P_V(z) \times \{d(z,V)\} \subset U ,$$

and consider the projections $\Pi_1(U) := \prod_{i=1}^{n} (a_i, b_i)$ and $\Pi_2(U) = (\alpha, \beta)$ onto $\mathbb{R}^n$ and $\mathbb{R}$ respectively. Since P_V is upper semicontinuous and $d(\cdot, V)$ continuous at z, we can find a neighborhood U_z of z such that for every $w \in U_z$

$$P_V(w) \subset \Pi_1(U) \quad \text{and} \quad \{d(w,V)\} \subset \Pi_2(U) .$$

Therefore $P_V(w) \times \{d(w,V)\} \subset U$, which shows that the mapping

$z \mapsto (P_V(z), d(z,V))$ is upper semicontinuous at z.

The converse is trivial.

Consider now the example 3.1 and the mappings p and C_z defined there and assume that V is a linear subspace of Z. It is clear that if $A : V \times \mathbb{R} \to Z \times \mathbb{R}$ is defined by $A(v,r) := (-v,-r)$ and $b := (-z,0)$ we have $C_z(u) \leq 0$ if and only if $A(u) \leq b$, where $u = (v,r)$.

Lemma 4.10. Consider the linear problem $MPI(\sigma_z)$ in 3.1 and let V be a non-empty proximinal subspace of Z. If for each $z \in Z$, $\dim P_V(z) < \infty$ and P_V is upper semicontinuous, then the condition (ii) in the theorem 4.5 is fulfilled for each parameter $b = (-z,0)$.

Proof. Let $U_b \subset L_{A,p}$ be a neighborhood of b. Since V is proximinal and $v \in P_V(z)$ if and only if P_b contains $(v, \|v - z\|)$, for each $b' \in U_b$, the set $P_{b'}$ is non-empty.

By choosing $\Pi(b')$ in $P_{b'}$, for each b' in U_b we define a mapping $\Pi : U_b \to U$ such that $\Pi(b') \in P_{b'}$, for every $b' \in U_b$.

On the other hand, since P_V is upper semicontinuous, by lemma 4.9 P_b is upper semicontinuous.

Now let $(b_n) \subset U_b$ and $b_n \to b$.

Since $\dim P_V(z) < \infty$ for all $z \in Z$, we have $\dim P_b < \infty$, for every $b \in Z \times \{0\}$.

Thus from Michael's theorem (see [5], page 180) $H := \bigcup_{n \in \mathbb{N}} P_{b_n} \cup P_b$ is compact.

Hence the subset $\{\Pi(b_n) : n \in \mathbb{N}\}$ of H is relatively compact, which concludes the proof.

As the Slater-condition is always fulfilled for the problem $BA(z)$, it follows from 4.5, 4.6 and 4.10 the

Corollary 4.11. Let Z be a real normed space, V be a non-empty proximinal subspace of Z with the properties:

(i) for each $z \in Z$ $\dim P_V(z) < \infty$;

(ii) the metric projection P_V is upper semicontinuous;

If for every $z \in Z$ with $0 \in P_V(z)$ there exists a w*-open convex set $U^* \subset Z^*$ such that

$$N_b \supset U^* \cap \text{ext}(B_{Z^*}) \supset \bigcap_{v \in P_V(z)} \{e^* \in \text{ext}(B_{Z^*}) \; ; \; e^*(z-v) = \|v-z\|\}$$

then P_V is lower semicontinuous.

The result above was proved by BROSOWSKI & WEGMANN in [3].

Corollary 4.12. If in the theorem 4.5 Y is a partially ordered Banach space, the conditions (i), (ii), (iii) (or the Slater-condition) and (iv) are fulfilled then the optimal set mapping P has a continuous selection.

Proof. The result is an immediate consequence of theorem 4.5 and Michael's selection theorem (see e.g. [4]).

REFERENCES

[1] Bank, B., Guddat, J., Klatte, D., Kummer, B., and Tammer, K., (1983) Non-linear Parametric Optimization (Birkhäuser Verlag, Basel-Boston, Stuttgart).

[2] Brosowski, B. (1982) Semi-infinite parametric optimization (Peter Lang Verlag, Frankfurt a.M.-Bern).

[3] Brosowski, B. and Wegmann, R. (1973) On the lower semicontinuity of the set-valued metric projection, J. Approx. Theory 8, 84-100.

[4] Holmes, R.B. (1975) Geometric functional analysis and its applications (Springer Verlag, New York - Heidelberg).

[5] Michael, M. (1951) Topologies on the spaces of the subsets, Trans. Amer. Math. Soc. 71, 151-182.

A. R. da Silva, Instituto de Matemática, Universidade Federal de Rio de Janeiro, Caixa Postal 68530, 21944 Rio de Janeiro, Brazil.

Current address: Johann Wolfgang Goethe Universität, Fachbereich Mathematik, Robert-Mayer-Strasse 6-10, 6 Frankfurt am Main, West Germany.

International Series of
Numerical Mathematics, Vol. 72

RATE OF CONVERGENCE OF THE METHOD OF ALTERNATING PROJECTIONS

Frank Deutsch

Department of Mathematics, The Pennsylvania State University, University Park, PA, U.S.A.

Abstract

A proof is given of a rate of convergence theorem for the method of alternating projections. The theorem had been announced earlier in [8] without proof.

1. Introduction

Let M_1 and M_2 be two closed subspaces of the Hilbert space X and let $P_1 = P_{M_1}$, $P_2 = P_{M_2}$, and $P = P_{M_1 \cap M_2}$ denote the corresponding orthogonal projections onto M_1, M_2, and $M_1 \cap M_2$ respectively. The method of alternating projections (= the alternating method of von Neumann = the method of Kacmarz) may be described as follows. Starting with any $x \in X$, define the sequence (x_n) inductively by

$$x_0 = x, \quad x_{2n-1} = P_2(x_{2n-2}), \quad x_{2n} = P_1(x_{2n-1}) \quad (n=1,2,\ldots).$$

VON NEUMANN [23] showed that the sequence (x_n) thus generated converges to $P(x)$. In particular, since

$x_{2n} = (P_1P_2)^n(x)$, he proved that

$$(\mathrm{MAP})_2 \qquad \lim_{n\to\infty} \|(P_1P_2)^n(x) - P(x)\| = 0$$

for every $x \in X$.

HALPERIN [15] extended the method to more than two subspaces. Thus we have

<u>1.1 Theorem</u> (Von Neumann-Halperin). Let M_i $(i=1,2,\ldots,k)$ be closed subspaces of the Hilbert space X, let $P_i = P_{M_i}$ denote the orthogonal projection onto M_i, and let P denote the orthogonal projection onto $\bigcap_1^k M_i$. Then

$$(\mathrm{MAP})_k \qquad \lim_{n\to\infty} \|(P_1P_2\ldots P_k)^n(x) - P(x)\| = 0$$

for every $x \in X$.

It is not hard to verify that Theorem 1.1 holds more generally when the M_i are closed linear varieties (i.e. translates of subspaces) with $\bigcap_1^k M_i \neq \phi$. Stated in this generality, Theorem 1.1 implies the "iterative method of Kacmarz" [18] for solving linear systems of equations.

It seems remarkable that the method of alternating projections is at the heart of so many areas of mathematics. An incomplete list might mention:

1. <u>Linear prediction theory</u> ([29]).
2. <u>Approximating multivariate functions by sums of univariate ones</u> ([13], [20], [19]).
3. <u>Abstract approximation theory</u> ([11], [2], [7]).

4. Sparse systems of linear equations ([10], [27]).

5. Relaxation methods for linear inequalities ([1], [21], [26]).

6. Multigrid methods for numerical solution of partial differential equations ([14], [3], [4]).

7. Least-change secant updates for quasi-Newton methods for solving functional equations ([5], [6]).

8. Computed tomography ([17], [25], [16]).

2. Rate of Convergence Theorem

While the method of alternating projections always converges, the rate of convergence may be arbitrarily slow [12]. However, if the angle between the subspaces is large, then the convergence is rapid. More precisely, SMITH, SOLMON, and WAGNER [25] have proved the following rate of convergence theorem.

2.1 Theorem (Smith, Solmon, Wagner) Let X, M_i, P_i and P be as in Theorem 1.1 and let θ_i denote the angle between the subspaces M_i and $\bigcap_{i+1}^{k} M_j$ for $i = 1,2,\ldots,k-1$. Then for every $x \in X$ and $n = 1,2,\ldots,$

$$\|(P_1P_2\ldots P_k)^n(x) - P(x)\| \le c^n\|x-P(x)\|, \tag{2.1.1}$$

where $c = \left[1 - \prod_{i=1}^{k-1} \sin^2\theta_i\right]^{\frac{1}{2}}$.

2.2 Definition. The angle between two subspaces M and N is defined to be the angle θ between 0 and $\pi/2$ whose cosine is given by the expression

$$(2.2.1)\quad c(M,N) := \sup\{|\langle x,y\rangle| \mid x \in M \cap (M \cap N)^{\perp},\ \|x\| \le 1,\ y \in N \cap (M \cap N)^{\perp},\ \|y\| \le 1\}.$$

That is, $\theta = \text{Arc}\cos[c(M,N)]$.

In the case $k = 2$, a rate of convergence theorem was established independently by FRANCHETTI and LIGHT [12] and the writer [8] (who were both unaware of Theorem 2.1 at the time). In [8], the error factor c^n in (2.1.1) was replaced by the sharper bound $c^{2n-1}(1-c^2)^{-1}$. Since this result was announced without proof in [8], it is the purpose of this note to supply the proof.

<u>2.3 Theorem</u>. Let M_1 and M_2 be closed subspaces of the Hilbert space X, let $P_i = P_{M_i}$ $(i=1,2)$, and $P = P_{M_1 \cap M_2}$. If $c := c(M_1,M_2) < 1$, then for each $x \in X$,

$$(2.3.1)\qquad \|(P_1P_2)^n(x) - P(x)\| \le \frac{c^{2n-1}}{1-c^2}\|x - P_1P_2(x)\|$$

for $n = 1,2,\ldots$. In particular, if $c(M_1,M_2) = 0$, then $P_1P_2 = P$ and the algorithm converges in one step.

<u>Proof</u>. Fix any $x \in X$ and define $x_0 = x$, $x_{2n-1} = P_2(x_{2n-2})$, $x_{2n} = P_1(x_{2n-1})$ $(n=1,2,\ldots)$. Clearly $x_{2n} = (P_1P_2)^n(x)$ $(n=0,1,\ldots)$. Since $PP_1 = P = PP_2$, we have that

$$P(x_{2n}) = P(P_1(x_{2n-1})) = P(x_{2n-1}) = P(P_2(x_{2n-2}))$$
$$= P(x_{2n-2}).$$

In particular,

$$(2.3.2) \qquad P(x_n - x_m) = 0 \qquad (n,m \geq 0).$$

Next, using the well-known orthogonality of the error $x - P_i(x)$ to M_i $(i=1,2)$, we see that

$$(2.3.3) \qquad \begin{aligned} x_{2n-1} - x_{2n} &= x_{2n-1} - P_1(x_{2n-1}) \in M_1^{\perp} \\ x_{2n} - x_{2n+1} &= x_{2n} - P_2(x_{2n}) \in M_2^{\perp} \end{aligned}$$

for all n. Using (2.3.2), (2.3.3), and the fact that $x_{2n} \in M_1$, $x_{2n-1} \in M_2$ for all $n \geq 1$, we get

$$\begin{aligned} \|x_{2n+2}-x_{2n}\|^2 &= \langle x_{2n+2}-x_{2n}, x_{2n+2}-x_{2n}\rangle \\ &= \langle(x_{2n+2}-x_{2n+1} + (x_{2n+1}-x_{2n-1}) + (x_{2n-1}-x_{2n}), \\ &\qquad x_{2n+2}-x_{2n}\rangle \\ &= \langle x_{2n+1}-x_{2n-1}, x_{2n+2}-x_{2n}\rangle \\ &= \langle x_{2n+1}-x_{2n-1}-P(x_{2n+1}-x_{2n-1}), x_{2n+2}-x_{2n}- \\ &\qquad P(x_{2n+2}-x_{2n})\rangle \\ &\leq c\|x_{2n+1}-x_{2n-1}-P(x_{2n+1}-x_{2n-1})\|\,\|x_{2n+2}-x_{2n}- \\ &\qquad P(x_{2n+2}-x_{2n})\| \\ &\leq c\|x_{2n+1}-x_{2n-1}\|\,\|x_{2n+2}-x_{2n}\|. \end{aligned}$$

Hence

$$(2.3.4) \qquad \|x_{2n+2}-x_{2n}\| \le c\|x_{2n+1}-x_{2n-1}\| \qquad (n=1,2,\ldots).$$

Similarly,

$$(2.3.5) \qquad \|x_{2n+1}-x_{2n-1}\| \le c\|x_{2n}-x_{2n-2}\| \qquad (n=2,3,\ldots)$$

and

$$(2.3.6) \qquad \|x_3-x_1\| \le \|x_2-x_0\|.$$

Combining (2.3.4) and (2.3.5), we obtain

$$(2.3.7) \qquad \|x_{2n+2}-x_{2n}\| \le c^2\|x_{2n}-x_{2n-2}\| \qquad (n=2,3,\ldots).$$

By induction, we deduce

$$(2.3.8) \qquad \|x_{2n+2}-x_{2n}\| \le (c^2)^{n-1}\|x_4-x_2\| \le c^{2n-1}\|x_3-x_1\| \le c^{2n-1}\|x_2-x_0\|.$$

By a repeated application of (2.3.8), it follows that for all positive integers p,

$$\|x_{2(n+p)}-x_{2n}\| \le \|x_{2(n+p)}-x_{2(n+p-1)}\| + \|x_{2(n+p-1)} - x_{2(n+p-2)}\| + \ldots + \|x_{2(n+1)} - x_{2n}\|$$

$$\leq [c^{2(n+p-1)-1} + c^{2(n+p-2)-1} + \ldots + c^{2n-1}] \|x_2 - x_0\|$$

$$\leq c^{-1} \sum_{k=n}^{\infty} c^{2k} \|x_2 - x_0\| = \frac{c^{2n-1}}{1-c^2} \|x_2 - x_0\| .$$

Letting $p \to \infty$ in this inequality and using Theorem 1.1, we obtain

$$\|P(x) - x_{2n}\| \leq \frac{c^{2n-1}}{1-c^2} \|x_2 - x_0\| \qquad (n=1,2,\ldots)$$

and this verifies (2.3.1). ∎

2.4 Remarks. (1) It is easy to verify that (in the case when $k = 2$) the error factor $c^{2n-1}(1-c^2)^{-1}$ in (2.3.1) is smaller than the error factor c^n in (2.1.1). Indeed, $c^{2n-1}(1-c^2)^{-1} = o(c^n)$. It would be interesting to know whether analogous sharper bounds are available for more than two subspaces (i.e. $k > 2$). I suspect so.

(2) For applications of Theorem 2.3, it is important to know when $c(M_1,M_2) < 1$ and $c(M_1,M_2) = 0$. These two situations (among other things) are characterized in the following lemma.

2.5 Lemma. Let M_1, M_2 be closed subspaces of the Hilbert space X. Then

(1) $0 \leq c(M_1,M_2) \leq 1$.

(2) $c(M_1,M_2) = c(M_2,M_1)$

(3) $c(M_1,M_2) = \|P_{M_1}P_{M_2} - P_{M_1 \cap M_2}\| = \|P_{M_1}P_{M_2}P_{M_1^\perp}\|$

(4) $c(M_1,M_2) < 1$ if and only if $M_1 \cap (M_1 \cap M_2)^{\perp} + M_2 \cap (M_1 \cap M_2)^{\perp}$ is closed.

(5) $c(M_1,M_2) = 0$ if and only if P_{M_1} and P_{M_2} commute: $P_{M_1}P_{M_2} = P_{M_2}P_{M_1}$. (In this case, $P_{M_1}P_{M_2} = P_{M_1 \cap M_2}$.)

<u>Proof</u>. (1) and (2) are obvious. To prove (3), first note that $x \in M_1 \cap (M_1 \cap M_2)^{\perp}$ iff $x \in M_1$ and $P_{M_1 \cap M_2}(x) = 0$. Similarly, $y \in M_2 \cap (M_1 \cap M_2)^{\perp}$ iff $y \in M_2$ and $P_{M_1 \cap M_2}(y) = 0$. Thus

$$
\begin{aligned}
c(M_1,M_2) &= \sup\{|\langle x-P_{M_1 \cap M_2}(x),\ y-P_{M_1 \cap M_2}(y)\rangle| \mid x \in M_1,\ \|x\| \le 1, \\
&\qquad\qquad y \in M_2,\ \|y\| \le 1\} \\
&= \sup\{|\langle P_{M_1}(x)-P_{M_1 \cap M_2}(P_{M_1}(x)),\ P_{M_2}(y)-P_{M_1 \cap M_2}(P_{M_2}(y))\rangle| \mid \\
&\qquad\qquad \|x\| \le 1,\ \|y\| \le 1\} \\
&= \sup\{|\langle (I-P_{M_1 \cap M_2})P_{M_1}(x),(I-P_{M_1 \cap M_2})P_{M_2}(y)\rangle| \mid \\
&\qquad\qquad \|x\| \le 1,\ \|y\| \le 1\} \\
&= \sup\{|\langle P_{M_2}(I-P_{M_1 \cap M_2})P_{M_1}(x),y\rangle| \mid \|x\| \le 1,\ \|y\| \le 1\} \\
&= \|P_{M_2}(I-P_{M_1 \cap M_2})P_{M_1}\| = \|P_{M_2}P_{M_1}-P_{M_1 \cap M_2}\| \\
&= \|P_{M_2}P_{M_1}(I-P_{M_1 \cap M_2})\| = \|P_{M_2}P_{M_1}P_{(M_1 \cap M_2)^{\perp}}\| \\
&= \|P_{M_2}P_{M_1}P_{\overline{M_1^{\perp}+M_2^{\perp}}}\| = \|P_{M_2}P_{M_1}P_{M_2^{\perp}}\|.
\end{aligned}
$$

(4) For notational simplicity, let $A = M_1 \cap (M_1 \cap M_2)^{\perp}$, $B = M_2 \cap (M_1 \cap M_2)^{\perp}$, and $\alpha = c(M_1, M_2)$. Then A and B are closed subspaces and we must show that $\alpha < 1$ iff $A + B$ is closed.

Assume first that $\alpha < 1$. By the definition of α we have

$$|\langle a,b\rangle| \le \alpha \|a\| \, \|b\| \tag{2.5.1}$$

for all $a \in A$, $b \in B$. Now let $c_n \in A + B$ and $c_n \to c$. We must show that $c \in A + B$. Write $c_n = a_n + b_n$. where $a_n \in A$, $b_n \in B$. Then $\{c_n\}$ is bounded so there exists a constant $\rho > 0$ such that

$$\begin{aligned}\rho &\ge \|c_n\|^2 = \|a_n+b_n\|^2 = \|a_n\|^2 + 2\mathrm{Re}\langle a_n,b_n\rangle + \|b_n\|^2 \\ &\ge \|a_n\|^2 - 2|\langle a_n,b_n\rangle| + \|b_n\|^2 \ge \|a_n\|^2 - 2\alpha\|a_n\|\,\|b_n\| + \|b_n\|^2 \\ &= (\|a_n\|-\|b_n\|)^2 + 2(1-\alpha)\|a_n\|\,\|b_n\| .\end{aligned}$$

Since $0 \le \alpha < 1$, it follows that both the sequences $((\|a_n\|-\|b_n\|)^2$ and $(\|a_n\|\,\|b_n\|)$ are bounded. From this it follows that both sequences $(\|a_n\|)$ and $(\|b_n\|)$ are bounded. Since X is reflexive, by passing to a subsequence if necessary, we may assume that both (a_n) converges weakly to some $a \in A$ and (b_n) converges weakly to some $b \in B$: $a_n \xrightarrow{w} a$, $b_n \xrightarrow{w} b$. Thus $c_n = a_n + b_n \xrightarrow{w} a + b$. But $c_n \to c$ so $c = a + b \in A + B$ and $A + B$ is closed.

Conversely, suppose $A + B$ is closed. Then $X' := A + B$ is a Hilbert space and $A \cap B = \{0\}$ implies that $X' = A \oplus B$ is the direct sum of A and B. By a well-known consequence of the open mapping theorem, the projection Q of X' onto A along B is continuous. If the result were false, then $\alpha = 1$ and there would exist $a_n \in A$, $b_n \in B$ with $\|a_n\| = 1 = \|b_n\|$ and

$$1 = \alpha = \lim_{n\to\infty} \langle a_n, b_n\rangle .$$

Then

$$\begin{aligned}\|a_n - b_n\|^2 &= \|a_n\|^2 - 2\mathrm{Re}\langle a_n, b_n\rangle + \|b_n\|^2 \\ &= 2 - 2\mathrm{Re}\langle a_n, b_n\rangle \to 0\end{aligned}$$

implies that

$$a_n = Q(a_n - b_n) \to Q(0) = 0$$

which is absurd. Thus $\alpha < 1$ and the proof of (4) is complete.

(5) If $c(M_1, M_2) = 0$ then by (3), $P_{M_1}P_{M_2} = P_{M_1\cap M_2}$ amd by (2), $P_{M_2}P_{M_1} = P_{M_1\cap M_2}$. Hence $P_{M_1}P_{M_2} = P_{M_2}P_{M_1} = P_{M_1\cap M_2}$.

Conversely, if $P_{M_1}P_{M_2} = P_{M_2}P_{M_1}$, then

$$P_{M_1}P_{M_2}P_{M_1^\perp} = P_{M_2}P_{M_1}P_{M_1^\perp} = P_{M_2} \cdot 0 = 0$$

and (3) implies that $c(M_1, M_2) = 0$. ■

3. References

1. S. Agmon, The relaxation method for linear inequalities, Canadian J. Math., 6(1954), 382-392.

2. B. Atlestam and F. Sullivan, Iteration with best approximation operators, Rev. Roum, Math. Pures et Appl., 21(1976), 125-131.

3. R.E. Bank and I. Dupont, An optimal order process for solving finite element equations, Math. Comput., 36(1981), 35-51.

4. D. Braess, The contraction number of a multigrid method for solving the Poisson equation, Numer. Math., 37(1981), 387-404.

5. J.E. Dennis, Jr. and R.B. Schabel, Least change secant updates for quasi-Newton methods, SIAM Review, 21(1979), 443-459.

6. J.E. Dennis, Jr. and H.F. Walker, Convergence theorems for least-change secant update methods, SIAM J. Numer. Anal., 18(1981), 949-987.

7. F. Deutsch, The alternating method of von Neumann, Multivariate Approximation Theory (edited by W. Schempp and K. Zeller), ISNM 51, Birkhäuser, Basel, 1979.

8. F. Deutsch, Von Neumann's alternating method: The rate of convergence, Approximation Theory IV (edited by C.K. Chui, L.L. Schumaker, and J.D. Ward), Academic Press, New York, 1983.

9. F. Deutsch, Applications of von Neumann's alternating projections algorithm, Proc. of Conference on "Mathematical Methods in Operations Research", Sofia, Bulgaria, 1983 (to appear).

10. E. Durand, Solutions Numeriques des Équations Algebriques, Tome II, Masson, Paris, 1961.

11. C. Franchetti, On the alternating approximation method, Sezione Sci., 7(1973), 169-175.

12. C. Franchetti and W. Light, On the von Neumann alternating algorithm in Hilbert Space, CAT Report #28, Nov. 1982.

13. M. Golomb, Approximation by functions of fewer variables, On Numberical Approximation (ed. R. Langer), Univ. of Wisconsin Press, Madison, 1959.

14. J.E. Gunn, The solution of difference equations by semi-explicit iterative techniques, SIAM J. Numer. Anal., 2(1965), 24-45.

15. I. Halperin, The product of projection operators, Acta Sci. Math. (Szeged),. 23(1962), 96-99.

16. C. Hamaker and D.C. Solmon, The angles between the null spaces of x-rays, J. Math. Anal. Appl., 62(1978), 1-23.

17. G.N. Hounsfield, Computerized transverse axial scanning (tomography): Part I Description of system, British J. Radiol., 46(1973), 1016-1022.

18. S. Kacmarz, Bull. intern. Acad. Polonaise des Sciences A(1937), 355-357.

19. C.T. Kelley, A note on the approximation of functions of several variables by sums of functions of one variable, J. Approx. Theory, 33(1981), 179-189.

20. W.A. Light and E.W. Cheney, On the approximation of a bivariate function by the sum of univariate functions, J. Approx. Theory, 29(1980), 305-322.

21. T.S. Motzkin and I.J. Schoenberg, The relaxation method for linear inequalities, Canadian J. Math., 6(1954), 393-404.

22. H. Nakano, Spectral Theory in the Hilbert Space, Tokyo, 1953.

23. J. von Neumann, Functional Operators--Vol. II. The Geometry of Orthogonal Spaces, Annals of Math. Studies #22, Princeton University Press, 1950. (This is a reprint of mimeographed lecture notes first distributed in 1933.)

24. M.J.D. Powell, A new algorithm for unconstrained optimization, Nonlinear Programming (edited by J.B. Rosen, O.L. Mangasarian, and K. Ritter), Academic Press, New York, 1970.

25. K.T. Smith, D.C. Solmon, and S.L. Wagner, Practical and mathematical aspects of the problem of reconstructing objects from radiographs, Bull. Amer. Math. Soc., 83(1977), 1227-1270.

26. J.E. Spingarn, A primal-dual projection method for solving systems of linear inequalities, Linear Algebra and Its applications, to appear.

27. K. Tanabe, Projection method for solving a singular system of linear equations and its applications, Numer. Math., 17(1971), 203-214.

28. N. Wiener, On the factorization of matrices, Comment. Math. Helv., 29(1955), 97-111.

29. N. Wiener and P. Masani, The prediction theory of multivariate stochastic processes, II: The linear predictor, Acta. Math., 93(1957), 95-137.

Prof. Frank Deutsch, Department of Mathematic, The Pennsylvania State University, University Park, Pennsylvania, 16802, U.S.A.

International Series of
Numerical Mathematics, Vol. 72

SINGULAR PERTURBATION IN LINEAR DIFFERENTIAL INCLUSIONS - CRITICAL CASE

A.L. Dontchev and V.M. Veliov

Institute of Mathematics, Sofia, Bulgaria

1. Introduction

In this paper we consider the following linear differential inclusion

$$(1)\qquad \begin{bmatrix} \dot{x} \\ \beta\dot{y} \end{bmatrix} \in A(t)\begin{bmatrix} x \\ y \end{bmatrix} + B(t)U(t)\,, \qquad \begin{bmatrix} x(0) \\ y(0) \end{bmatrix} = \begin{bmatrix} x^0 \\ y^0 \end{bmatrix}$$

$$x(t) \in R^n,\ y(t) \in R^m,\ U(t) \in R^r,\ t \in [0,T]\,,$$

where β is a small positive scalar parameter; the initial conditions (x^0,y^0) and the interval $[0,T]$ are given. We say that for fixed $\beta > 0$, $z_\beta(\cdot) = (x_\beta(\cdot),y_\beta(\cdot))$ is a solution of (1) iff $z_\beta(\cdot)$ is absolutely continuous and fulfilles (1) for almost all $t \in [0,T]$. In the case considered this is equivalent to the following: there exists an integrable selection $u_\beta(\cdot)$ of $U(\cdot)$ on $[0,T]$ such that $(x_\beta(\cdot),y_\beta(\cdot))$ solves the linear system

$$(2a)\qquad \dot{x} = A_1(t)x + A_2(t)y + B_1(t)u_\beta(t)\,, \quad x(0) = x^0,$$

$$(2b)\qquad \beta\dot{y} = A_3(t)x + A_4(t)y + B_2(t)u_\beta(t)\,, \quad y(0) = y^0,$$

where

$$A(t) = \begin{bmatrix} A_1(t) & A_2(t) \\ A_3(t) & A_4(t) \end{bmatrix}, \quad B(t) = \begin{bmatrix} B_1(t) \\ B_2(t) \end{bmatrix}$$

and the matrices $A_i(t)$, $B_j(t)$ have appropriate dimensions. The equations (2) can be interpreted as a control system model with state (x,y) and control $u(t) \in U(t)$, where β represents small physical parameters. If we set $\beta = 0$, the inclusion (1) degenerates to

$$(3) \qquad \begin{bmatrix} \dot{x} \\ 0 \end{bmatrix} \in A(t) \begin{bmatrix} x \\ y \end{bmatrix} + B(t)U(t) ,$$

that is, the differential equation (2b) becomes algebraic one

$$(4) \qquad 0 = A_3(t)x + A_4(t)y + B_2(t)u(t)$$

and the y-part of the solution may be not longer absolutely continuous. The perturbation represented by the parameter β leads to a change of the solution space, therefore this perturbation is often called singular.

The singular perturbation analysis finds numerous applications in design and optimization of control systems, for surveys see Vasileva and Dmitriev (1982) and Kokotovic (1984).

In this paper we present a continuation of our previous works concerning the continuity of the multivalued mapping: "singular parameter $\beta \longrightarrow$ set of trajectories of (1)" at $\beta = 0$. In Dontchev and Veliov (1983) and Dontchev (1983) we assumed that the eigenvalues $\lambda(A_4(t))$ of the matrix $A_4(t)$ have strictly negative real parts for all $t \in [0,T]$, $\mathrm{Re}\,\lambda(A_4(t)) < 0$. For this case we obtain that the x-part of the set of solutions is pointwise Hausdorff continuous but the entire set is pointwise lower semicontinuous only. Dontchev and Veliov (1984) developes these results for the case when

$$A_4(t) = \begin{bmatrix} A_{41}(t) & \varphi(\beta)A_{42}(t) \\ \varphi(\beta)A_{43}(t) & A_{44}(t) \end{bmatrix}$$

where Re $\lambda(A_{41}(t)) < 0$, Re $\lambda(A_{44}(t)) > 0$ and $\varphi(\beta) \to 0$ as $\beta \to 0$. Here we consider the so-called critical case, when the eigenvalues of $A_4(t)$ may have zero real parts.

In the sequel we use the following basic condition

A1. The matrices $A_1(t)$, $A_2(t)$ and $B_1(t)$ have integrable components and $A_3(t)$, $A_4(t)$ and $B_2(t)$ have $L_\infty[0,T]$ components. The matrix $A_4(t)$ is invertible for all $t \in [0,T]$. The components of the matrix

$$D(t) = A_2(t)A_4^{-1}(t)$$

are absolutely continuous and the components of $\dot{D}(t)A_4^{-1}(t)$ have bounded variations on $[0,T]$. The multivalued mapping U(.) is compact valued and upper semicontinuous on $[0,T]$. If $Y_\beta(t,s)$ denotes the fundamental matrix solution of the equation

$$\beta\dot{y} = A_4(t)y$$

normalized at $t = s$, then there exists a constant $C > 0$ independent of β, t and s such that

$$|Y_\beta(t,s)| \leqslant C$$

for all $\beta > 0$, $0 \leqslant s \leqslant t \leqslant T$.

On this assumption we prove that the x-part $X(\beta)$ of the solutions of (1) is lower semicontinuous in L_p-weak topology for every $1 \leqslant p < +\infty$, and pointwise lower semicontinuous. If U(.) is convex valued than $X(\beta)$ is L_p-weakly upper semicontinuous, as well.

In Section 3 we suppose that A1 and the following condition hold:

A2. The matrix $A_4(t)$ is constant, $A_4(t) = A_4$. The matrices $D(t)$, $B_0(t) = D(t)B_1(t)$ and $B_2(t)$ are continuous on $[0,T]$. The multivalued mapping $U(t)$ is Hausdorff continuous on $[0,T]$. The matrix $\exp(A_4 t)$ is periodic with period $\omega > 0$.

We prove that $X(\beta)$ possesses a pointwise Hausdorff limit when $\beta \longrightarrow 0$. We find this limit and show that it may be larger than $X(0)$, i.e. $X(\beta)$ is essentially pointwise lower semicontinuous.

As an application of these results, in Section 4 we discuss the order reduction of control systems in critical case.

Throughout the paper $|\cdot|$ designates the euclidean norm. The norm in $L_p[0,T]$ will be $\|\cdot\|_p$ and the uniform norm is $\|\cdot\|_C$. By $O(\beta)$ we denote a (vector) function which tends to zero with β.

2. L_p weak continuity

Assume that A1 holds. We are interested in the x-part of the inclusion (1), hence by solving (4) with respect to y, one can rewrite the reduced inclusion (3) as

$$(5) \qquad \dot{x} \in A_0(t)x + B_0(t)U(t) \;, \quad x(0) = x^0,$$

where

$$A_0(t) = A_1(t) - D(t)A_3(t) \;,$$

$$B_0(t) = B_1(t) - D(t)B_2(t) \;.$$

For fixed $\beta \geqslant 0$ denote

$$X(\beta) = \{x(\cdot),\ (x(\cdot),y(\cdot)) \text{ solves } (1)\} \;.$$

Then $X(0)$ will be the set of solutions of (5). We start with the following technical lemma:

Lemma 1. Let $x_\beta(\cdot) \in X(\beta)$ and $u_\beta(\cdot)$ be the corresponding control function according to (2ab). Suppose that $\tilde{x}_\beta(\cdot)$ solves

$$x(t) = x^0 + \int_0^t (A_0(s)x(s) + (B_0(s) + D(t)Y_\beta(t,s)B_2(s))u_\beta(s))ds.$$

Then

$$\lim_{\beta \to 0} \| x_\beta - \tilde{x}_\beta \|_C = 0 .$$

Proof. We show first that $X(\beta)$ is bounded in $C[0,T]$ uniformly in β. Using Cauchy formula and integrating by parts we obtain consequently

$$x_\beta(t) = x^0 + \int_0^t A_1(s)x_\beta(s)ds + \int_0^t A_2(s)Y_\beta(s,0)y^0 ds$$

$$+ \frac{1}{\beta}\int_0^t A_2(s) \int_0^s Y_\beta(s,\tau)(A_3(\tau)x_\beta(\tau) + B_2(\tau)u_\beta(\tau))d\tau ds$$

$$+ \int_0^t B_1(s)u_\beta(s)ds$$

$$= x^0 + \int_0^t A_1(s)x_\beta(s)ds + \int_0^t A_2(s)Y_\beta(s,0)y^0 ds$$

$$+ \int_0^t\int_\tau^t D(s)\frac{\partial}{\partial s}Y_\beta(s,\tau)ds(A_3(\tau)x_\beta(\tau) + B_2(\tau)u_\beta(\tau))d\tau$$

$$+ \int_0^t B_1(s)u_\beta(s)ds$$

$$= x^0 + \int_0^t A_1(s)x_\beta(s)ds + \int_0^t A_2(s)Y_\beta(s,0)y^0 ds$$

$$+ \int_0^t (D(t)Y_\beta(t,\tau) - D(\tau) - \int_\tau^t \dot{D}(s)Y_\beta(s,\tau)ds$$

$$\times (A_3(\tau)x_\beta(\tau) + B_2(\tau)u_\beta(\tau))d\tau + \int_0^t B_1(s)u_\beta(s)ds. \tag{6}$$

Using A1 and applying Gronwall lemma one simply gets that $x_\beta(.)$ is uniformly bounded. Since $x_\beta(.)$ is arbitrarily chosen from $X(\beta)$ then $X(\beta)$ is uniformly bounded.

By the last equality in (6) in order to complete the proof one should show that

$$I_1^\beta(t) = \int_0^t A_2(s)Y_\beta(s,0)y^0 ds = O(\beta) \quad ,$$

$$I_2^\beta(t) = \int_0^t D(t)Y_\beta(t,s)A_3(s)x_\beta(s)ds = O(\beta) \, ,$$

$$I_3^\beta(t) = \int_0^t \int_\tau^t D(s)Y_\beta(s,\tau)ds(A_2(\tau)x_\beta(\tau) + B_2(\tau)u_\beta(\tau))d\tau = O(\beta).$$

The first relation follows immediately by an integration by parts.

Let $\delta > 0$ be arbitrarily chosen. One can find a matrix $A^{\delta}(t)$ with C^1 components such that

$$\| A^{\delta} - A_4^{-1}A_3\|_1 < \delta .$$

Then

$$I_2^{\beta}(t) = \int_0^t D(t)Y_{\beta}(t,s)A_4(s)(A_4^{-1}(s)A_3(s) - A^{\delta}(s))x_{\beta}(s) + \bar{I}_2^{\beta}(t)$$

where

$$\bar{I}_2^{\beta}(t) = \int_0^t D(t)Y_{\beta}(t,s)A_4(s)A^{\delta}(s)x_{\beta}(s)ds$$

$$= - \beta \int_0^t D(t)\frac{\partial}{\partial s}Y_{\beta}(t,s)A^{\delta}(s)x_{\beta}(s)ds$$

$$= - \beta D(t)A^{\delta}(t)x_{\beta}(t) + \beta D(t)A^{\delta}(0)x^{0}$$

$$+ \beta \int_0^t D(t)Y_{\beta}(t,s)(\dot{A}^{\delta}(s)x_{\beta}(s) + A^{\delta}(s)(A_1(s)x_{\beta}(s)$$

$$+ B_1(s)u_{\beta}(s)))ds .$$

The uniform boundedness of $x_{\beta}(.)$ and $u_{\beta}(.)$ and A1 implies that

$$|\bar{I}_2^{\beta}(t)| \leqslant c(\delta)O(\beta)$$

uniformly in $t \in [0,T]$, hence

$$|I_2^{\beta}(t)| \leqslant c_1\delta + c(\delta)O(\beta) .$$

or

$$\lim_{\beta \to 0} |I_2^\beta(t)| \leqslant c_1 \delta ,$$

which means

$$I_2^\beta(t) = O(\beta) .$$

Integrating by parts the inner integral in $I_3^\beta(t)$ and using the uniform boundedness of $x_\beta(.)$ and $u_\beta(.)$ we complete the proof.

Theorem 1. For every p, $1 \leqslant p < +\infty$ the multivalued mapping X(.) is lower semicontinuous in the weak $L_p[0,T]$ topology. If U(.) is convex valued then X(.) is upper semicontinuous in the weak $L_p[0,T]$ topology.

Proof. Take $x_o(.) \in X(0)$ and let $u_o(.)$ correspond to $x_o(.)$, that is $x_o(.)$ solves

$$(7) \qquad \dot{x} = A_o(t)x + B_o(t)u(t) , \quad x(0) = x^0$$

for $u(.) = u_o(.)$. Let $(x_\beta(.), y_\beta(.))$ solve (1) for the selection $u_o(.)$, i.e. $(x_\beta(.), y_\beta(.))$ solve (2ab) for $u = u_o(.)$. Taking into account Lemma 1 it is sufficient to prove that the function

$$\varphi_\beta(t) = \int_0^t D(t)Y_\beta(t,s)B_2(s)u_o(s)ds$$

converges L_p-weakly to zero. We show that this is true for a more general situation, namely when

$$\varphi_\beta(t) = \int_0^t D(t)Y_\beta(t,s)B_2(s)u_\beta(s)ds ,$$

where $u_\beta(t) \in U(t)$ for a.e. $t \in [0,T]$ and $\beta > 0$. It is classically known that the L_p-weak convergence of $\varphi_\beta(.)$ is equivalent to

$$\|\varphi_\beta\|_p \text{ is bounded as } \beta \longrightarrow 0 ;$$

$$\int_0^t \varphi_\beta(s)ds \longrightarrow 0 \quad \text{as} \quad \beta \longrightarrow 0$$

for all $t \in [0,T]$. The first condition is trivially satisfied from A1. In order to obtain the second condition

$$\int_0^t \varphi_\beta(s)ds = \int_0^t \int_s^t D(\tau)Y_\beta(\tau,s)d\tau B_2(s)u_\beta(s)ds \longrightarrow 0 \text{ as } \beta \longrightarrow 0$$

it is sufficient to use an intgration by parts in the inner integral (the inner integral is the same as in $I_3^\beta(t)$ in the proof of Lemma 1). Hence $\varphi_\beta(.)$ converges L_p-weakly to zero.

Let $x_\beta(.) \in X(\beta)$ with corresponding selection (control) $u_\beta(.)$ and $x_\beta(.)$ converges L_p-weakly to some $x(.)$. By Lemma 1

$$x_\beta(t) = \tilde{x}_\beta(t) + O(\beta)$$

and, moreover

$$\tilde{x}_\beta(t) = \bar{x}_\beta(t) + \varphi_\beta(t),$$

where $\bar{x}_\beta(.)$ solves (7) for $u(.) = u_\beta(.)$, i.e. $\bar{x}_\beta(.) \in X(0)$. If we prove that $X(0)$ is L_p weakly closed then the proof will be complete.

By a standart argument the set

$$W = \{u(.),\ u(t) \in U(t) \text{ for a.e. } t \in [0,T],\ u(.) - \text{measurable}\}$$

is compact in the $L_2[0,T]$ weak topology. Let $F(t,s)$ be the fundamental matrix solution of the equation

$$\dot{x} = A_0(t)x$$

normalized at $t = s$. Since the operator

$$K : L_2[0,T] \longrightarrow L_p[0,T]$$

$$(Ku)(t) = \int_0^t F(t,s)B_0(s)u(s)ds$$

is weakly continuous, and $X(0) = KW$ then $X(0)$ is L_p weakly closed.

3. Pointwise lower semicontinuity and pointwise limit

We continue the analysis of the multivalued mapping $X(.)$ assuming that A1 holds. Let us denote as $S(t,\beta)$ the set of values of $X(\beta)$ at t and let $S(t,0)$ correspond to $X(0)$:

$$S(t,\beta) = \{x,\ x = x_\beta(t),\ x_\beta(.) \in X(\beta)\} .$$

By Lyapunov's vector measure theorem both $S(t,\beta)$ and $S(t,0)$ are compact and convex and remain the same if we replace $U(t)$ by $\mathrm{co}\, U(t)$.

Theorem 2. Let $t^o \in (0,T]$ be fixed. For every $x \in S(t^o,0)$ and for every $\beta > 0$ there exists $x_\beta \in S(t,\beta)$ such that $x_\beta \longrightarrow x$ as $\beta \longrightarrow 0$, i.e. $X(\beta)$ is pointwise lower semicontinuous at $\beta = 0$.

Proof. Let the function $u(.)$ correspond to x, $u(t) \in U(t)$, $t \in [0,t^o]$. Take $u(.)$ in (2ab) and let $x_\beta(.)$ be the corresponding solution. By Lemma 1 it is sufficient to show that

$$\lim_{\beta \to 0} \int_0^{t^0} D(t)Y_\beta(t^0,s)B_2(s)u(s)ds = 0 \; .$$

This follows immediately from the fact that $Y_\beta(t^0,s)v$ converges L_2-weakly to zero for every vector $v \in R^n$. Really,

$$\int_0^{t^0} Y_\beta(t^0,s)vds = -\beta \int_0^{t^0} \frac{\partial}{\partial s} Y_\beta(t^0,s)A_4^{-1}(s)vds$$

and integrating by parts as in the proof of Lemma 1,we obtain the desired result.

In the sequel we assume that A1 and A2 hold. We show that for every $t \in [0,T]$ $S(t,\beta)$ possesses a Hausdorff limit which, however, may be larger than $S(t,0)$.

For fixed $t \in [0,T]$ introduce the set

$$S(t) = \int_0^t F(T,T - s)R(t - s)ds \; , \tag{8}$$

where $R(.)$ is a continuous compact and convex valued multivalued mapping defined as

$$R(s) = \frac{1}{\omega} \int_0^{\omega} (B_0(s) + D(s)\exp(A_4\tau)B_2(s))U(s)d\tau \; . \tag{9}$$

(We remind that $\omega > 0$ is the period of $\exp(A_4t)$).

<u>Theorem 3</u>. For every $t^0 \in [0,T]$

$$S(t^0,\beta) \longrightarrow S(t^0) \quad \text{as} \quad \beta \longrightarrow 0$$

in the sense of Hausdorff.

<u>Proof</u>. Let $\delta > 0$ be fixed. We prove that: (A) $S(t^o)$ is contained in a δ-neighbourhood of $S(t^o,\beta)$ for all sufficiently small $\beta > 0$, and (B) $S(t^o,\beta)$ is contained in a δ-neighbourhood of $S(t^o)$ for all sufficiently small $\beta > 0$.

(A) Let $x \in S(t^o)$ then there exists a measurable function $z(.)$, $z(t) \in R(t^o - t)$ such that

$$x = \int_0^{t^o} F(t^o,t^o - t)z(t)dt \cdot$$

Furthermore

$$x = \sum_{k=0}^{\left[\frac{t^o}{\omega\beta}\right]-1} F(t^o,t^o-k\omega\beta) \int_{k\omega\beta}^{(k+1)\omega\beta} z(t)dt + O(\beta)$$

$$\in \sum_{k=0}^{\left[\frac{t^o}{\omega\beta}\right]-1} F(t^o,t^o-k\omega\beta) \int_{k\omega\beta}^{(k+1)\omega\beta} R(t^o-t)dt + O(\beta)$$

$$= \omega\beta \sum_{k=0}^{\left[\frac{t^o}{\omega\beta}\right]-1} F(t^o,t^o-k\omega\beta)R(t^o-k\omega\beta) + O(\beta),$$

where $O(\beta) \longrightarrow 0$ as $\beta \longrightarrow 0$ for the fixed t^o. Here and further we use the following standard assertion:

(T) Let $P(.)$ and $Q(.)$ be two compact and convex valued Hausdorff continuous multivalued mappings on $[t_1,t_2]$ such that the Hausdorff distance $d(P(t),Q(t)) \leqslant \delta$ for some $\delta > 0$ and for all $t \in [t_1,t_2]$. Then for every measurable selection $u(.)$, $u(t) \in P(t)$ there exists a measurable $v(.)$, $v(t) \in Q(t)$ such that $\|u - v\|_\infty \leqslant \delta$.

Then the last equality follows from the convexity and

continuity of R(.). Hence, there exist measurable functions $u_k^\beta(\cdot)$, $u_k^\beta(t) \in U(t^o - k\omega\beta)$ such that

$$x = \beta \sum_{k=0}^{\left[\frac{t^o}{\omega\beta}\right]-1} F(t^o, t^o - k\omega\beta) \int_0^\omega (B_0(t^o - k\omega\beta)$$

$$+ D(t^o - k\omega\beta)\exp(A_4 s)B_2(t^o - k\omega\beta))u_k^\beta(s)ds + O(\beta)$$

$$= \beta \sum_{k=0}^{\left[\frac{t^o}{\omega\beta}\right]-1} \int_0^\omega F(t^o, t^o - k\omega\beta - s\beta)(B_0(t^o - k\omega\beta - s\beta)$$

$$+ D(t^o - k\omega\beta - s\beta)\exp(A_4(s+k\omega))B_2(t^o - k\omega\beta - s\beta))u_k^\beta(s)ds$$

$$+ O(\beta).$$

Choosing new independent variable $\tau = t^o - (k\omega + \beta)s$ we get

$$x = \sum_{k=0}^{\left[\frac{t^o}{\omega\beta}\right]-1} \int_{t^o-(k+1)\omega\beta}^{t^o-k\omega\beta} F(t^o, \tau)(B_0(\tau)$$

$$+ D(\tau)\exp(A_4 \frac{t^o - \tau}{\beta})B_2(\tau))u_k^\beta(\frac{t^o - \tau}{\beta} - k\omega)d\tau + O(\beta) .$$

Introduce the function

$$u^\beta(\tau) = \begin{cases} u_k^\beta(\frac{t^o - \tau}{\beta} - k\omega) , & \tau \in \left[t^o - (k+1)\omega\beta, t^o - k\omega\beta\right) \\ u , & \left[0, t^o - (\left[\frac{t^o}{\omega\beta}\right] - 1)\omega\beta\right) , \end{cases}$$

where $u \in U(0)$ is arbitrarily chosen. Then

$$(10)\qquad x = \int_0^{t^o} F(t^o,\tau)(B_o(\tau) + D(\tau)\exp(A_4\frac{t^o-\tau}{\beta})B_2(\tau))u^\beta(\tau)d\tau + O(\beta)$$

Since $u^\beta(t) \in U(t - k\omega\beta)$ for $t \in [t^o-(k+1)\omega\beta, t^o-k\omega\beta[$ then from the continuity of co U(.) and from (T) (taking co U(t) instead of U(t) does not change $S(t^o,\beta)$) it follows that there exists a measurable function $\bar{u}^\beta(.)$, $\bar{u}^\beta(t) \in$ co U(t) such that

$$\|\bar{u}^\beta - u^\beta\|_\infty < \frac{\delta}{2} \quad .$$

Then, denoting as B_n the unique ball in R^n from (10) we obtain

$$x \in \int_0^{t^o} F(t^o,\tau)(B_o(\tau) + D(\tau)\exp(A_4\frac{t^o-\tau}{\beta})B_2(\tau))\bar{u}^\beta(\tau)d\tau$$

$$+ O(\beta) + \frac{\delta}{2}B_n \subset S(t^o,\beta) + O(\beta) + \frac{\delta}{2}B_n \ ,$$

where $O(\beta) \longrightarrow 0$ uniformly in $x \in S(t^o)$. This gives us part A of the proof. Bart B uses the same argument as A but in the reverse order. Let $x_\beta \in S(t^o,\beta)$ and $\bar{u}_\beta(.)$ be the corresponding selection of U(.) such that

$$x_\beta = \int_0^{t^o} F(t^o,t)(B_o(t) + D(t)\exp(A_4\frac{t^o-t}{\beta})B_2(t))\bar{u}_\beta(t)dt$$

Then

$$x_\beta = \sum_{k=0}^{\left[\frac{t^o}{\omega\beta}\right]-1} \int_{t^o-(k+1)\omega\beta}^{t^o-k\omega\beta} F(t^o,t)(B_o(t)$$

$$+ D(t)\exp(A_4 \frac{t^o-t}{\beta})B_2(t))\bar{u}_\beta(t)dt + O(\beta) .$$

From (T) there exists $u_\beta(.)$ - measurable, $u_\beta(t) \in \mathrm{co}\,U(t^o-k\omega\beta)$ for $t \in [t^o-(k+1)\omega\beta, t^o-k\omega\beta)$, $k = 0,\dots,[T/\omega\beta]-1$, such that

$$\|u_\beta - \bar{u}_\beta\|_\infty < \frac{\delta}{2}$$

for all sufficiently small β. Taking $\tau = t^o - (k\omega + s)\beta$ we obtain

$$x_\beta \in \beta \sum_{k=0}^{\left[\frac{t^o}{\omega\beta}\right]-1} \int_0^\omega F(t^o, t^o-k\omega\beta-\beta s)(B_o(t^o-k\omega\beta-\beta s)$$

$$+ D(t^o-k\omega\beta)\exp(A_4(s+k\omega))B_2(t^o-k\omega\beta-\beta s))u_\beta(t^o-k\omega\beta-\beta s))ds$$

$$+ O(\beta) + \frac{\delta}{2}B_n$$

Since $u_\beta(t^o-k\omega\beta-\beta s) \in U(t^o-k\omega\beta)$ for $s \in [0,\omega]$ then

$$\frac{1}{\omega}\int_0^\omega F(t^o, t^o-k\omega\beta)(B_o(t^o-k\omega\beta)$$

$$+ D(t^o-k\omega\beta)\exp(A_4 s)B_2(t^o-k\omega\beta))u_\beta(t^o-k\omega\beta-\beta s)ds$$

$$\subset R(t^o-k\omega\beta) .$$

From the continuity of $F(t^o,..)$, $B_o(.)$, $D(.)$ and $B_2(.)$ and the periodicity of $\exp(A_4 s)$ we conclude that

$$x_\beta \in \beta \sum_{k=0}^{\left[\frac{t^o}{\omega\beta}\right]-1} F(t^o, t^o-k\omega\beta)R(t^o-k\omega\beta) + O(\beta) + \frac{\delta}{2}B_n .$$

By (T) one can find a measurable function $z_\beta(\cdot)$, $z_\beta(t) \in R(t)$, $t \in (0,T]$ such that

$$\left| x_\beta - \int_0^T F(t^o, t^o-t) z_\beta(t^o-t)dt \right| \leqslant O(\beta) + \frac{3}{4}\delta ,$$

which proves part B of the proof.

Clearly by theorems 2 and 3

$$S(t,0) \subset S(t) .$$

The following example shows that $S(t)$ may be larger than $S(t,0)$:

Example. Consider the inclusion

$$\begin{aligned} \dot{x} &= y_2 , & x(0) &= 0 , \\ \beta\dot{y}_1 &= y_2 , & y_1(0) &= 0 , \\ \beta\dot{y}_2 &\in -y_1 + [-1,1] , & y_2(0) &= 0 . \end{aligned}$$

Here the reduced inclusion is

$$\dot{x} \in \{0\} \quad \text{and } S(t,0) = \{0\}$$

for all $t \geqslant 0$. By simple computations we get

$$S(t) = \left[-\frac{2t}{\pi} , \frac{2t}{\pi}\right]$$

for all $t \geqslant 0$.

4. Conclusions

The results obtained in the previous two sections can be directly applied to justify the order reduction of linear control systems in critical case. Clearly, taking $\beta = 0$ in (1) or (2ab) may lead to an essential simplification of the model. This, however, may be accompanied by discontinuities of the system performance.

Theorems 1 and 3 shows that the choise of the reduced model in the critical case will depend on the topology we would like to work in. For example, if we are interested in <u>trajectory optimization</u>, that is, our optimization problem has the form

$$\int_0^T V(x(t),u(t),t)dt \longrightarrow \min$$

subject to (2ab), then, under some continuity conditions for $V(.)$ the optimal value will depend continuously on β at $\beta = 0$. Then the reduced model will have the standard form

$$\dot{x} = A_0(t)x + B_0(t)u \; .$$

However, if the performance index depends on the state at some instant of time, say

$$g(x(T)) \longrightarrow \min$$

then, in order to get continuity of the optimal value one should use the following system as a reduced model

$$\dot{x} = A_0(t)x + w(t) \; , \quad x(0) = x^0 \; , \; w(t) \in R(t) \; ,$$

where $R(t)$ is given by (9) or equivalently it is the reachable set at ω of the system

$$\omega \frac{d}{ds} z = (B_0(t) + A_2(t)A_4^{-1}(t)\exp(A_4 s)B_2(t))u(s),$$

$$z(0) = 0,$$

$$u(s) \in U(s) .$$

REFERENCES

Dontchev A.L., Veliov V.M.(1983) Singular perturbation in Mayer's problem for linear systems.SIAM J.Control and Optim. 21,566-581.

Dontchev A.L., Veliov V.M. (1984) Singular perturbations in linear control systems with weakly coupled stable and unstable fast subsystems,to be published in J.Math.Anal.Appl.

Dontchev A.L.(1983) Perturbations,approximations and sensitivity analysis of optimal control systems.Lecture Notes in Control and Inf.Sc. 52 (Springer,Berlin).

Kokotovic P.(1984) Applications of singular perturbation techniques to control problems, to appear in SIAM Review.

Vasileva A.B.,Dmitriev M.G. (1982) Singular perturbations in optimal control problems, Itogi Nauki i Technili (VINITI, Moscow),in Russian.

Institute of Mathematics,Bulgarian Academy of Sciences,
1090 Sofia,P.O.Box 373,Bulgaria.

International Series of
Numerical Mathematics, Vol. 72

SENSITIVITY ANALYSIS IN GENERALIZED RATIONAL APPROXIMATION WITH RESTRICTED DENOMINATOR

by

Jacob Flachs

National Research Institute for Mathematical Sciences,
CSIR, P O Box 395, Pretoria 0001, S.A.

Abstract

Generalized rational approximation with restricted denominator is viewed as a parametric program in which all the given functions that are in= volved, as well as the compact space on which they are defined, play the role of parameters. We prove a general theorem from which we derive the Lipschitz behaviour in a certain sense of the optimal value and the solution set.

1. Introduction

The subject of this paper is the sensitivity analysis of generalized rational approximation problems with restricted denominator. Under certain assumptions, we shall establish a local Lipschitz property of the optimal value and the Lipschitz behaviour in a certain sense of the solution set. The ap= proach encompasses the behaviour of both the optimal value and the solution set under noise and under various perturbations (including discretizations) of the compact set on which the involved functions are defined, these aspects being important in practice.

We denote by $\mathcal{B}[X_1,X_2]$ the linear space of all continuous linear ope= rators from the normed space X_1 to the normed space X_2. Let T be a (sequen= tially) compact space and denote

$$A(T) \triangleq C(T) \times C(T) \times B[\mathbb{R}^n,C(T)] \times B[\mathbb{R}^m,C(T)] \times B[\mathbb{R}^m,C(T)].$$

The elements $A \in A(T)$ are thus quintuples $A = (f,r,P,Q,R)$ with $f,r \in C(T)$, $P \in B[\mathbb{R}^n,C(T)]$ and $Q,R \in B[\mathbb{R}^m,C(T)]$.

Let $K \subset \mathbb{R}^m$ be a fixed convex set. For any $A = (f,r,P,Q,R) \in A(T)$ (i.e., $f,r \in C(T)$, $P \in B[\mathbb{R}^n,C(T)]$, $Q,R \in B[\mathbb{R}^m,C(T)]$) and any compact subset $V \subset T$ consider the following parametric program (with A and V as parameters):

$$\mathbf{P}(A,V) \quad \alpha(A,V) \triangleq \inf_{x,y} \{\|f - \frac{Px}{Qy}\|_V \mid y \in C(A,V) \cap K\},$$

with its solution set denoted by $S(A,V)$, where $C(A,V) \triangleq \bigcap_{t \in V} \{y \mid Q(t)y \geq 1,\ R(t)y \geq r(t)\}$ and $\|\cdot\|_V$ is the maximum norm in $C(V)$. This notation needs further explanation: denoting $Q_i = Qe_i^m$ where e_i^m is the i-th unit vector in $\mathbb{R}^m$, it follows that $Qy = \sum_{i=1}^m Q_i y_i \in C(T)$ for any $y = (y_1,\ldots,y_m)^T \in \mathbb{R}^m$. The functions Q_i are called the columns of Q, while for each $t \in T$ the row vector $(Q_1(t),\ldots,Q_m(t))$ is denoted by $Q(t)$. The same applies for R and P : $R(t) = (R_1(t),\ldots,R_m(t))$ and $P(t) = (P_1(t),\ldots,P_n(t))$ where $R_i = Re_i^m \in C(T)$ and $P_i = Pe_i^n \in C(T)$ are called the columns of R and P respectively. Thus, for any compact subset $V \subset T$ and any quintuple $A = (f,r,P,Q,R) \in A(T)$, $\mathbf{P}(A,V)$ is a (generalized) rational approximation where the variable y is restricted to a supplementary constraint determined by (fixed) K and linear inequalities involving R and r.

Our problem $\mathbf{P}(A,T)$ includes the cases considered by Dunham (ref [5]), Krabs (ref [9]), where $K \triangleq \mathbb{R}^m$, $R = -Q$ (r is constant in [9]); it also includes the problem considered by Dunham (ref [3]) and Kaufman and Taylor (ref [8]) where $K \triangleq \{(y_1,\ldots,y_m)^T \mid |y_i| \leq 1,\ i = 1,\ldots,m\}$ and R and r do not appear (equivalently $R \equiv 0$, $r \equiv -1$).

There are a number of papers dealing with the continuity properties of the rational approximation with restricted denominator. In ref. [3] and

[9], these properties are shown for converging sequences $V^k \subset T$ (A is kept fixed), while in ref [4] continuity is proved for the particular case where f^k converges to f (other parameters are kept fixed).

We shall use the following additional notations:
For $A = (f,r,P,Q,R) \in \mathcal{A}(T)$, the functions $f,r,P_i,Q_j,R_j \in C(T)$ are called the columns of A; the maximum norm is considered for P,Q,R and A (viewed as an operator), i.e.,

$$\|P\| \triangleq \max_{i=1,\dots,n} \{\|P_i\|\}; \quad \|Q\| \triangleq \max_{j=1,\dots,m} \{\|Q_j\|\}; \quad \|R\| \triangleq \max_{j=1,\dots,m} \{\|R_j\|\};$$

$$\|A\| \triangleq \max\{\|f\|, \|r\|, \|P\|, \|Q\|, \|R\|\},$$

where $\|f\|$, $\|r\|$, $\|P_i\|$, $\|Q_j\|$ and $\|R_j\|$ are the maximum norms over T of the respective functions.

For each two subsets U and Z of a metric space with metric ρ we define

$$\rho[U,Z] \triangleq \sup_{u\in U} \inf_{z\in Z} \rho(u,z) \; ; \quad \omega[U;Z] \triangleq \max \{\rho[U,Z], \; \rho[Z,U]\};$$

and if U and Z are subsets of a normed space, we define

$$d[U,Z] \triangleq \sup_{u\in U} \inf_{z\in Z} \|u-z\| \; ; \quad \delta[U,Z] \triangleq \max \{d[U,Z], d[Z,U]\}.$$

We shall also use the set $\mathring{\mathcal{C}}(A,V)$ defined by

$$\mathring{\mathcal{C}}(A,V) \triangleq \bigcap_{t\in V} \{y \mid Q(t)y > 1, \quad R(t)y > r(t)\}.$$

Our first result is the following: If $\mathring{\mathcal{C}}(A,T) \cap K$ is non-void and bounded, then $\alpha(A,T)$ is a locally Lipschitz function in A.
Then we consider the case where (A,V) varies. From a more general theorem the following is derived: If the columns of A are Lipschitz functions, $\mathring{\mathcal{C}}(A,V) \cap K$ is non-void and bounded and rank $P|_V = n$, then

(a) there are $\varepsilon,\sigma,M > 0$ such that for any $A' \in \mathcal{A}(T)$ with $\|A'-A\| \le \varepsilon$ and any compact subset $V' \subset T$ with $\omega[V',V] \leq \sigma$, the following inequality holds:

$$|\alpha(A',V') - \alpha(A,V)| \leq M\|A'-A\| + M\omega[V',V];$$

(b) if in addition, K is polyhedral or $\mathbb{R}^m$, then for each T_0 and V_0 belonging to certain families (of subsets of V) there are $\varepsilon,\sigma,M > 0$ such that for any $A' \in A(T)$ with $\|A'-A\| \leq \varepsilon$ and any compact subset $V' \subset T$ with $\omega[V',V] \leq \sigma$, the following holds:

$$d[S(A',V'),\ S(A,T_0 \cup V_0)] \leq M\|A'-A\| + M\omega[V',V].$$

Finally, we show that (b) implies the upper semi-continuity of the solution set $S(\cdot,\cdot)$.

Our result regarding the behaviour of $\alpha(\cdot,T)$ is derived using Robin=son's theorems (ref [10]), while that regarding the solution set $S(\cdot,\cdot)$ is proved using Hoffman's theorem (ref [2]). For the general case our results are essentially based on a reduction theorem, which is presented in Section 2.

Note that $S(A,T) \neq \phi$ if $C(A,T) \cap K$ is compact. This can be proved using standard compactness and lower semi-continuity arguments. Although in almost all theorems we suppose the above compactness condition, the reduction theorem (Proposition 1) does not require it. Moreover, Proposition 1 is true even for the case where **P**(A,T) has no solution (i.e. $S(A,T) = \phi$).

2. A reduction theorem

In this section we show the existence of a discretization of V (i.e. a finite subset $V_0 \subset V$) such that $\alpha(A,V_0) = \alpha(A,V)$ and discuss properties of $\mathbf{P}(A,V_0)$ and its solution set $S(A,V_0)$.

For each $A \in A(T)$, each compact subset $V \subset T$ and each positive in=teger ℓ we define the following family $\Sigma^\ell(A,V)$:

$$\Sigma^\ell(A,V) \triangleq \{V_0 \mid V_0 \subset V,\ 1 \leq \text{card } V_0 \leq \ell,\ \alpha(A,V_0) = \alpha(A,V)\}.$$

First we make the following remark:

Remark 1. Let $\{G_\tau\}$ and $\{g_\tau\}$ be two arbitrary collections of m-dimentional vectors and numbers respectively. Suppose that there is $y' \in R^m$ such that $G_\tau^T y' > g_\tau$ for any τ. Then,

$$\text{int}[\cap_\tau \{y \mid G_\tau^T y \geq g_\tau\}] \subset \cap_\tau \{y \mid G_\tau^T y > g_\tau\}.$$

Proof. Let $z \in \text{int}[\cap_\tau \{y | G_\tau^T y \geq g_\tau\}]$ and suppose that $G_\tau^T z = g_\tau$ for some τ. Considering points of the form $z + \varepsilon G_\tau$ for small $|\varepsilon|$, from our choice of z it follows that $G_\tau = 0$ and $g_\tau = 0$, which contradicts our assumption. □

Note that the following theorem does not need a solution of **P**(A,T).

Proposition 1. Let T be a compact space, $K \subset \mathbb{R}^m$ be a convex set and let $A = (f,r,P,Q,R) \in A(T)$. Suppose that $\mathring{C}(A,T) \cap K \neq \phi$ and denote $m_0 \triangleq \dim K$. Then,

(i) $\Sigma^{n+m_0+1}(A,T) \neq \phi$

(ii) for any compact subset $T'' \subset T$ and any compact subset $T' \subset T$ that includes a set belonging to $\Sigma^\ell(A,T)$ (for some ℓ),

$$\alpha(A,T'') \leq \alpha(A,T) = \alpha(A,T') \text{ and } S(A,T) \subset S(A,T').$$

Proof. Let $b \in \mathbb{R}^m$ be such that $\mathcal{L} \triangleq \text{aff } K - b$ is a m_0 dimensional subspace of $\mathbb{R}^m$; let $M : \mathcal{L} \to \mathbb{R}^{m_0}$ be an isomorhpism and denote

$$\bar{Q} \triangleq QM^{-1},\ \bar{R} \triangleq RM^{-1},\ K_0 \triangleq M(K-b),\ g \triangleq Qb,\ s \triangleq Rb$$

$$C \triangleq \bigcap_{t\in T} \{z|\bar{Q}(t)z + g(t) \geq 1,\ \bar{R}(t)z + s(t) \geq r(t)\}$$

$$\mathring{C} \triangleq \bigcap_{t\in T} \{z|\bar{Q}(t)z + g(t) > 1,\ \bar{R}(t)z + s(t) > r(t)\}.$$

It is easy to see that by the transformation $z = M(y-b)$, the problem **P**(A,T) is equivalent to

$$\alpha(A,T) = \inf_{x,z} \{\| f - \frac{Px}{\bar{Q}z+g} \|_T \mid z \in C \cap K_0\}. \tag{1}$$

Our supposition implies that $\mathring{C} \neq \phi$, which, together with Remark 1 and the compactness of T, yields

$$\text{int } C = \mathring{C} = \text{int } \mathring{C}. \tag{2}$$

Since K is a convex set, M(relint K-b) = int K_0 which, by our supposition and (2), implies

$$\text{int}[C \cap K_0] = \mathring{C} \cap \text{int } K_0 \neq \phi. \tag{3}$$

Now, consider the function

$$E(x,z;t) \triangleq \begin{cases} \left| f(t) - \dfrac{P(t)x}{\bar{Q}(t)z+g(t)} \right|, & z \in \text{int } K_0,\ \bar{Q}(t)z+g(t) > 1, \bar{R}(t)z+s(t) > r(t) \\ \infty & \text{otherwise.} \end{cases}$$

Clearly, from (1)-(3) it follows that

$$\alpha(A,T) = \inf_{x,z} \max_{t \in T} E(x,z;t),$$

which implies

$$\bigcap_{t \in T} \{(x,z) \mid E(x,z;t) < \alpha(A,T)\} = \phi.$$

It is easy to see that the function E, together with $\alpha(A,T)$, satisfies properties A1-A3 of ref [6]. The proof of this is similar to that of ref [6]. Thus, by Lemma 1 of ref [6], there is a finite subset $T_0 = (t_1,\ldots,t_k) \subset T$ such that $k \leq n + m_0 + 1$ and

$$\bigcap_{t \in T_0} \{(x,z) \mid E(x,z;t) < \alpha(A,T)\} = \phi.$$

It follows that

$$\alpha(A,T) \leq \inf_{x,z} \max_{t \in T_0} E(x,z;t) = \alpha(A,T_0). \tag{4}$$

On the other hand, denoting

$$\bar{E}(x,y;t) \triangleq \begin{cases} |f(t) - \frac{P(t)x}{Q(t)y}|, & y \in K, \quad Q(t)y \geq 1, \quad R(t)y \geq r(t) \\ \infty & \text{otherwise,} \end{cases}$$

we deduce that for any compact subsets $T' \subset T'' \subset T$ and any x,y,

$$\max_{t \in T'} \bar{E}(x,y;t) \leq \max_{t \in T''} \bar{E}(x,y;t) \leq \max_{t \in T} \bar{E}(x,y;t) \tag{5}$$

and that the problems **P**(A,T'), **P**(A,T") and **P**(A,T) are obtained by taking the infinum by (x,y) in each of the expressions of (5). Clearly, the inequa= lity of part (ii) follows, which, together with (4), implies that $T_0 \in \Sigma^{n+m_0+1}(A,T)$. Finally, invoking (5), the inclusion statement of (ii) is proved. □

Note that Proposition 1 remains true even when T is replaced by any compact subset $V \subset T$, i.e., if $\mathring{C}(A,V) \cap K \neq \phi$, then (i) $\Sigma^{n+m_0+1}(A,V) = \phi$; and (ii) for any compact subset $V'' \subset V$ and any compact subset $V' \subset V$ that includes a set of $\Sigma^{\ell}(A,V)$ (for some ℓ),

$$\alpha(A,V'') \leq \alpha(A,V) = \alpha(A,V') \text{ and } S(A,V) \subset S(A,V').$$

For each $A = (f,r,P,Q,R) \in A(T)$ and each compact subset $V \subset T$ we define the family $\Delta(A,V)$ of all <u>finite</u> subsets V_0 of V satisfying the following conditions:

(1) rank $P|_{V_0} = n$ where $P|_{V_0}$ is the operator P restricted to V_0 (i.e., the colums of $P|_{V_0}$ are the functions P_i restricted to V_0)

(2) the set $C(A,V_0) \cap K$ is bounded.

A sufficient condition for $\Delta(A,V) \neq \phi$ is given after the following lemma:

<u>Lemma 1</u>. Let $\{K^{\tau}\}$ be an arbitrary collection of closed convex sets in R^m. If $\bigcap_{\tau} K^{\tau}$ is non-void and bounded, then there are $\tau_1,\ldots,\tau_k$ such that

the set $\bigcap_{i=1}^{k} K^{\tau_i}$ is bounded.

Proof. Denote by ∂B the boundary of the unit ball in $\mathfrak{R}^m$ and by $0^+\mathcal{H}$ the recession cone of any given convex set $\mathcal{H}$. By our assumption, $0^+\bigcap_{\tau} K^\tau = \phi$ (see ref [11]) and thus,

$$\phi = \partial B \cap [0^+\bigcap_{\tau} K^\tau] = \partial B \cap \bigcap_{\tau} 0^+ K^\tau.$$

Since $0^+ K^\tau$ are closed sets and ∂B is compact, there are $\tau_1,\dots,\tau_k$ such that

$$\phi = \partial B \cap \bigcap_{i=1}^{k} 0^+ K^{\tau_i} = \partial B \cap [0^+ \bigcap_{i=1}^{k} K^{\tau_i}].$$

which implies that $0^+[\bigcap_{i=1}^{k} K^{\tau_i}] = \{0\}$. Thus, $\bigcap_{i=1}^{k} K^{\tau_i}$ is bounded (see ref [11]).

$\square$

Corollary 1. Let T be a compact space, $K \subset \mathfrak{R}^m$ be a closed convex set, and let $A = (f,r,P,Q,R) \in A(T)$ and a compact subset $V \subset T$. Suppose that (a) the set $\mathcal{\overset{\circ}{C}}(A,V) \cap K$ is non-void and bounded; and (b) rank $P|_V = n$. Then, $\Delta(A,V) \neq \phi$.

Proof. The compactness of V and assumption (a), together with Remark 1, imply int $C(A,V) = \mathcal{\overset{\circ}{C}}(A,V)$, and thus, by convexity and our assumption, the set $C(A,T) \cap K$ is bounded. Recalling Lemma 1, we deduce the existence of a finite subset $V' \subset V$ such that $C(A,V') \cap K$ is bounded. By assumption (b) there is a subset $V'' \subset V$ such that card $V'' = n$ and rank $P|_{V''} = n$ (see ref [1]). Remarking that $V' \cup V'' \in \Delta(A,V)$, we complete the proof. $\square$

3. Lipschitz properties for perturbed rational approximation

In this section we deal with the sensitivity analysis of the program $P(A,V)$, first when A veries and $V = T$, second - as a consequence of the first case and Proposition 1 - when both parameters A and V vary. We discuss the Lipschitz behaviour in a certain sense, of the solution set $S(A,V)$.

In our development we need the following two additional lemmas:

Lemma 2. Let T be a compact space, $K \subset \mathbb{R}^m$ be a closed convex set, and let $A = (f,r,P,Q,R) \in A(T)$. We further assume that a number $L > 0$ is given and for each $\varepsilon > 0$ denote $V(\varepsilon) \triangleq \{A' | A' \in A(T), \|A'-A\| \leq \varepsilon\}$. Then,

(i) if rank P = n then there is $\varepsilon > 0$ such that the point-to-set map $\{x \mid \|P'x\| \leq L\}$ (depending on P' and thus on A') is uniformly bounded on the set $V(\varepsilon)$;

(ii) if $\mathring{C}(A,T) \cap K$ is non-void and bounded, then there is $\varepsilon > 0$ such that the point-to-set map $C(A',T) \cap K$ is uniformly bounded and $\mathring{C}(A',T) \cap K \neq \phi$ on the set $V(\varepsilon)$.

Proof. (i) Since rank P = n, it follows that $\bigcap_{t \in T} \{x \,||\, P(t)x| \leq L\}$ is bounded. Remarking that $0 \in \bigcap_{t \in T} \{x \,||\, P(t)x| < L\}$, this is a particular case of (ii).

(ii) Let $y^* \in K$ with $y^* \in \mathring{C}(A,T)$. By the compactness of T there is $\eta > 0$ such that $Qy^* \geq 1 + \eta$ and $Ry^* \geq r + \eta$. Denoting $\varepsilon' = \min \{\frac{\eta}{4}, \frac{\eta}{4\|y^*\|}\}$, where $\|y^*\|$ is the ℓ_1 norm of y^*, it is easily seen that $\|A'-A\| \leq \varepsilon'$ implies $y^* \in \mathring{C}(A',T)$, which proves the second assertion of (ii). The uniform boundedness is proved as follows: By the compactness of T, our assumption and Remark 1 it follows that int $C(A,T) = \mathring{C}(A,T)$. Thus, by con= vexity and our assumption, $C(A,T) \cap K$ is bounded; let N be a bound thereof. Suppose that $C(\cdot,T) \cap K$ is not uniformly bounded around A. Thus, there is a sequence A^k converging in norm to A and also a sequence y^k with $y^k \in C(A^k,T) \cap K$ and $\|y^k\| \to \infty$. Choose $\lambda^k \in [0,1]$ such that for any large enough k,

$$z^k = \lambda^k y^* + (1-\lambda^k)y^k, \; N + 1 \leq \|z^k\| \leq N + 2.$$

Let z be an accumulation point of z^k. Thus $\|z\| \geq N + 1$. On the other hand, as we have already shown above, $y^* \in \mathring{C}(A^k,T) \subset C(A^k,T)$ for any large enough k;

thus, by the convexity of $C(A^k,T)$, it follows that $z^k \in C(A^k,T)$ for any large enough k. The point-to-set map $C(\cdot,T)$ being closed at A (see ref [7]), we deduce that $z \in C(A,T) \cap K$, which contradicts the boundedness of $C(A,T) \cap K$ by N. □

Lemma 3. Let $D \subset \mathcal{R}^n$, $\mathcal{E} \subset \mathcal{R}^m$ and let A be a subset of a normed space. We further assume that a function $\varphi : D \times \mathcal{E} \times A \to \mathcal{R}$ and a point-to-set map $\psi : A \to 2^{\mathcal{E}}$ are given and consider the following parametric program:

$$p(A) \qquad \bar{\alpha}(A) \triangleq \inf\{\varphi(x,y,A) \,|\, y \in \psi(A),\ x \in D\}.$$

Suppose that for any $A \in A$, $\psi(A)$ is a closed subset of $\mathcal{R}^m$ and that there is $M \geq 0$ such that for any $A',A'' \in A$ there is a solution (x',y') of $p(A')$ such that for any $y \in \mathcal{E}$:

$$|\varphi(x',y,A'') - \varphi(x',y',A')| \leq M\|y-y'\| + M\|A''-A'\|.$$

Then for any $A',A'' \in A$,

$$|\bar{\alpha}(A') - \bar{\alpha}(A'')| \leq M\|A'-A''\| + M\delta[\psi(A'), \psi(A'')].$$

where δ is defined in introduction.

Proof. Let $A',A'' \in A$ and let (x',y') and (x'',y'') be solutions of $p(A')$ and $p(A'')$ respectively. Without restricting the generality, we suppose that $\varphi(x',y',A') \leq \varphi(x'',y'',A'')$. Let $y \in \psi(A'')$ such that

$$\|y'-y\| = d[y',\psi(A'')] \leq \delta[\psi(A'),\psi(A'')]. \qquad (6)$$

Thus, $\varphi(x'',y'',A'') \leq \varphi(x',y,A'')$ and

$$0 \leq \varphi(x'',y'',A'') - \varphi(x',y',A') = \varphi(x'',y'',A'') - \varphi(x',y,A'') + \varphi(x',y,A'')$$
$$- \varphi(x',y',A') \leq \varphi(x',y,A'') - \varphi(x',y',A').$$

which, together with our assumption and (6), completes the proof. □

<u>Theorem 1</u>. Let T be a compact space, $K \subset \mathbb{R}^m$ be a closed convex set and let $A = (f,r,P,Q,R) \in \mathcal{A}(T)$. Suppose that (a) the set $\mathring{C}(A,T) \cap K$ is non-void and bounded; and (b) rank P = n. Then there are $\varepsilon, M > 0$ such that for any $A', A'' \in \mathcal{A}(T)$ satisfying $\|A'-A\|, \|A''-A\| \le \varepsilon$ the following holds:

$$|\alpha(A',T) - \alpha(A'',T)| \le M\|A'-A''\|.$$

<u>Proof</u>. For each $\varepsilon > 0$ denote

$$V(\varepsilon) \triangleq \{A' \mid A' = (f',r',P',Q',R') \in \mathcal{A}(T),\ \|A'-A\| \le \varepsilon\}.$$

Since by Lemma 2, there are ε', $N > 0$ such that the point-to-set map $C(\cdot,T) \cap K$ is uniformly bounded by N on the set $V(\varepsilon')$, assumption (a), together with Robinson's theorems of ref[10], implies the existence of $\varepsilon_0, M_0 > 0$ (with $\varepsilon_0 \le \varepsilon'$) such that for any A', $A'' \in V(\varepsilon_0)$,

$$\delta[C(A',T) \cap K,\ C(A'',T) \cap K] \le M_0\|A'-A''\|. \qquad (7)$$

Denote

$$\varepsilon_1 \triangleq \min\left\{\varepsilon_0, \frac{1}{4N}, \frac{1}{4\|Q\|M_0}, \frac{1}{\sqrt{2M_0}}\right\}$$

and also, by B, the ℓ_1-norm unit ball in $\mathbb{R}^m$. Let $y' \in C(A,T) \cap K + \varepsilon_1 M_0 B$; clearly, $y' = y + \varepsilon_1 M_0 b$, where $\|b\| \le 1$ and $y \in C(A,T) \cap K$; thus $\|y\| \le N$. Let $A' = (f',r',P',Q',R') \in V(\varepsilon_1)$; we have

$$Q'y' = Qy + \varepsilon_1 M_0 Qb + (Q'-Q)y + \varepsilon_1 M_0 (Q'-Q)b \ge$$

$$1 - \varepsilon_1 M_0\|Q\| - \|Q'-Q\|\|y\| - \varepsilon_1 M_0\|Q'-Q\| \ge \tfrac{1}{4}.$$

Thus,

$$[A' = (f',r',P',Q',R') \in V(\varepsilon_1) \text{ and } y' \in C(A,T) \cap K + \varepsilon_1 M_0 B] \Rightarrow Q'y' \ge \tfrac{1}{4}. \qquad (8)$$

For any $A' = (f',r',P',Q',R') \in \mathcal{A}(T)$ each solution (x',y') of $\mathbf{P}(A',T)$ satis= fies the following at any $t \in T$:

$$\left|\frac{P'(t)x}{Q'(t)y}\right| - |f'(t)| \le \left|f'(t) - \frac{P'(t)x'}{Q'(t)y'}\right| \le \alpha(A',T) \le \|f'\|$$

$$y' \in \mathcal{C}(A',T) \cap \mathcal{K}, \tag{9}$$

which implies

$$\|P'x'\| \le 2\|f'\|\|Q'y'\| \le 2\|f'\|\|Q'\|\|y'\|. \tag{10}$$

Since $\mathcal{C}(\cdot,T) \cap \mathcal{K}$ is uniformly bounded around A, relations (9) and (10), to= gether with Lemma 2 and (7), imply the existence of ε_2, $L > 0$ with $\varepsilon_2 \le \varepsilon_1$ such that the set

$$[\Pi \triangleq \bigcup_{\|P'-P\| \le \varepsilon_2} \{x \mid \|P'x\| \le L\}] \times [\mathcal{C}(A,T) \cap \mathcal{K} + \varepsilon_1 M_0 \mathcal{B}] \tag{11}$$

is bounded and contains all the solutions of **P**(A',T) for any $A' \in \mathcal{V}(\varepsilon_2)$. Moreover, by (8), the derivatives with respect to y, f(t), P(t) and Q(t) of $f(t) - P(t)x/Q(t)y$ are uniformly bounded on the set

$$T \times \Pi \times [\mathcal{C}(A,T) \cap \mathcal{K} + \varepsilon_1 M_0 \mathcal{B}] \times \mathcal{V}(\varepsilon_2).$$

Deducing that the conditions of Lemma 3 are satisfied for

$$\varphi(x,y,A) \triangleq \max_{t \in T} \left|f(t) - \frac{P(t)x}{Q(t)y}\right| , \ \psi(\cdot) \triangleq \mathcal{C}(\cdot,T) \cap \mathcal{K}$$

$$D \triangleq \Pi, \ \mathcal{E} \triangleq \mathcal{C}(A,T) \cap \mathcal{K} + \varepsilon_1 M_0 \mathcal{B}, \ \mathcal{A} \triangleq \mathcal{V}(\varepsilon_2),$$

it follows that there is an $M_1 > 0$ such that for any $A',A'' \in \mathcal{V}(\varepsilon_2)$,

$$|\alpha(A',T) - \alpha(A'',T)| \le M_1\|A'-A''\| + M_1\delta[\mathcal{C}(A',T) \cap \mathcal{K}, \mathcal{C}(A'',T) \cap \mathcal{K}],$$

which, together with (7), implies (i). □

In order to discuss the general case we need the following notations: For any $A = (f,r,P,Q,R) \in \mathcal{A}(T)$ and any finite sequence $V = \{v_1,\dots,v_q\} \subset T$ denote by A(V) the $q \times (2+n+2m)$ matrix whose i-th row is $A(v_i)$. This means that

$$A(V) \triangleq (f(V),r(V),P(V),Q(V),R(V)) \triangleq [f(v_i),r(v_i),P(v_i), Q(v_i), R(v_i)]_{i=1}^q.$$

Note that by this notation we have: $f(V) = [f(v_i)]_{i=1}^q$, $r(V) = [r(v_i)]_{i=1}^q$, $P(V) = [P(v_i)]_{i=1}^q$, $Q(V) = [Q(v_i)]_{i=1}^q$ and $R(V) = [R(v_i)]_{i=1}^q$.

$$\bar{A}(\varepsilon) \triangleq \left\{ A'(T',V') \;\middle|\; \begin{array}{l} T' = \{t_1',\ldots,t_q'\} \subset T,\ \rho(t_i,t_i') \le \sigma(\varepsilon),\ i=1,\ldots,q \\ \\ V' = \{v_1',\ldots,v_\ell'\} \subset T,\ A' \in A(T),\ \|A'-A\| \le \varepsilon \end{array} \right\}$$

For any matrix $A'(T',V') \in \bar{A}(\varepsilon)$ we have

$$\|A'(T')-A(T_0)\| \le \|A'(T')-A(T')\| + \|A(T')-A(T_0)\| \le \|A'-A\| + \varepsilon \le 2\varepsilon, \qquad (13)$$

the last two inequalities holding by (12) and the definition of $\bar{A}(\varepsilon)$; in the case where $\ell > 0$ (i.e., $V' \neq \phi$) we have

$$\|A'(V')\| \le \|A'\| \le \|A'-A\| + \|A\| \le \varepsilon + \|A\|,$$

which, together with (13) and the equality $\|A'(T',V')\| = \max\{\|A'(T')\|, \|A'(V')\|\}$, implies that for any $\varepsilon > 0$ the set $\bar{A}(\varepsilon)$ is bounded. Furthermore, by our assumption (a) and the compactness of T, we deduce the existence of $y^* \in K$ and $\eta > 0$ such that $Qy^* \ge 1 + \eta$ and $Ry^* \ge r + \eta$. We can easily choose an $\varepsilon' > 0$ such that for any $A' = (f',r',P',Q',R') \in A(T)$ with $\|A'-A\| \le \varepsilon'$ we have $Q'y^* \ge 1 + \eta/2$ and $R'y^* \ge r' + \eta/2$. Recalling the definition of $\bar{A}(\varepsilon')$, it follows that for any matrix $A^* \in cl\ \bar{A}(\varepsilon')$, $y^* \in C(A^*,I) \cap K$ where $I \triangleq \{1,\ldots,q+\ell\}$. Moreover, since $T_0 \in \Delta(A,T)$ means that $C(A(T_0),\{1,\ldots,q\}) \cap K = C(A,T_0) \cap K$ is bounded and rank $P(T_0) = n$, from (14) and Lemma 2 we deduce the existence of $\varepsilon'' > 0$ such that the sets $C(A^*,I) \cap K$ and $\{x\ \|P^*x\| \le L\}$ are uni= formly bounded on $\{A^* \mid A^* \in cl\ \bar{A}(\varepsilon'')\}$. Denote $\varepsilon_0 \triangleq \min\{\varepsilon',\varepsilon''\}$ and $\sigma_0 \triangleq \sigma(\varepsilon_0)$. Thus, in particular, the sets $C(A'(T',V'),I) \cap K = C(A',T' \cup V') \cap K$ and $\{x \mid \|P'(T',V')x\| \le L\}$ are uniformly bounded on $\bar{A}(\varepsilon_0) \subset \bar{A}(\varepsilon'')$, which, by the definition of $\bar{A}(\varepsilon_0)$ and W, trivially implies (i). In order to prove (ii) we remark that the boundedness of $\{x \mid \|P^*x\| \le L\}$ implies that rank P $= n$, and thus the conditions of Theorem 1 (in which $T = I = \{1,\ldots,q+\ell\}$) are fulfilled at any matrix $A \in cl\ \bar{A}(\varepsilon_0)$. Therefore $\alpha(\cdot,I)$ is locally Lipschitz on $cl\ \bar{A}(\varepsilon_0)$. This set being compact, it follows that $\alpha(\cdot,I)$ is a Lipschitz function; hence, there is $M > 0$ such that for any $A'(T',V')$, $A''(T'',V'') \in \bar{A}(\varepsilon_0)$,

For any $A = (f,r,P,Q,R) \in A(T)$ and any two finite sequences $V = \{v_1,\dots,v_q\}$ and $S = \{s_1,\dots,s_\ell\}$ define the following $(q+\ell) \times (2+n+2m)$ matrix

$$A(V,S) \triangleq (f(V,S),\ r(V,S),\ P(V,S),\ Q(V,S),\ R(V,S)) \triangleq$$

$$\left(\begin{bmatrix} f(V) \\ f(S) \end{bmatrix}, \begin{bmatrix} r(V) \\ r(S) \end{bmatrix}, \begin{bmatrix} P(V) \\ P(S) \end{bmatrix}, \begin{bmatrix} Q(V) \\ Q(S) \end{bmatrix}, \begin{bmatrix} P(V) \\ P(S) \end{bmatrix}\right).$$

We also remark that for any $A = (f,r,P,Q,R) \in A(T)$ and any finite sequence $V = \{v_1,\dots,v_p\} \subset T$: $C(A,V) = C(A(V), \{1,\dots,p\})$, the programs $\mathbf{P}(A,V)$ and $\mathbf{P}(A(V), \{1,\dots,p\})$ are identical.

Proposition 2. Let T be a compact metric space with metric ρ, $K \subset \mathbb{R}^m$ be a closed convex set and let $A = (f,r,P,Q,R) \in A(T)$. Suppose that (a) $\mathring{C}(A,T) \cap K$ is non-void and bounded; and (b) rank $P = n$. We further assume as given $T_0 = \{t_1,\dots,t_q\} \in \Delta(A,T)$, a number $L > 0$ and an integer $\ell \geq 0$. Then, there are ε_0, σ_0, $M > 0$ such that

(i) the sets $C(A',T' \cup V') \cap K$ and $\{x \mid \|P'(T',V')\| \leq L\}$ (depending on A', T' and V') are uniformly bounded on the parameter set

$$W \triangleq \left\{ (A',T',V') \;\middle|\; \begin{array}{l} A' = (f',r',P',Q',R') \in A(T),\ \|A'-A\| \leq \varepsilon_0 \\ T' = \{t_1',\dots t_q'\} \subset T,\ \rho(t_i,t_i') \leq \sigma_0,\ i=1,\dots,q \\ V' = \{v_1',\dots,v_\ell'\} \subset T \end{array} \right\}$$

(ii) $|\alpha(A',T' \cup V') - \alpha(A'',T'' \cup V'')| \leq M \max\{\|A'(T') - A''(T'')\|, \|A'(V') - A''(V'')\|\}$ for any $(A',T',V'), (A'',T'',V'') \in W$.

Proof. Since the columns of A are continuous functions, it follows that for any $\varepsilon > 0$ there is $\sigma(\varepsilon) > 0$ such that for any $S = \{s_1,\dots,s_q\} \subset T$ with $\rho(s_i,t_i) \leq \sigma(\varepsilon)$ $i=1,\dots,q$,

$$\|A(S) - A(T_0)\| \leq \varepsilon. \tag{12}$$

For each $\varepsilon > 0$ consider the following set $\bar{A}(\varepsilon)$ of $(q+\ell) \times (2+n+2m)$ matrices:

$$|\alpha(A'(T',V'),I) - \alpha(A''(T'',V''),I)| \le M\|A'(T',V') - A''(T'',V'')\|$$

$$= M \max\{\|A'(T')-A''(T'')\|, \|A'(V')-A''(V'')\|\}. \qquad (14)$$

Finally, since the programs $\mathbf{P}(A'(T',V'),I)$ and $\mathbf{P}(A''(T'',V''),I)$ are identical to the programs $\mathbf{P}(A',T' \cup V')$ and $\mathbf{P}(A'',T'' \cup V'')$ respectively, it follows that $\alpha(A'(T',V'),I) = \alpha(A',T' \cup V')$ and $\alpha(A''(T'',V''),I) = \alpha(A'',T'' \cup V'')$, which, together with (14), completes the proof. □

In the following general theorems we illustrate the behaviour of the optimal value $\alpha(A,V)$ and the solution set $\mathcal{S}(A,V)$ when the parameters A and V are perturbed.

<u>Theorem 2.</u> Let T be a compact metric space with metric ρ, $K \subset \mathbb{R}^m$ be a closed convex set, $V \subset T$ be a compact subset and let $A = (f,r,P,Q,R) \in \mathcal{A}(T)$. Suppose that (a) $\mathring{\mathcal{C}}(A,V)$ is non-void and bounded; and (b) rank $P|_V = n$. Denote $\ell = n + \dim K+1$.

Then for any $T_0 = \{t_1,\dots,t_q\} \in \Delta(A,V)$ there are ε, σ_0,σ_1, $M > 0$ such that for any $A',A'' \in \mathcal{A}(T)$ satisfying $\|A'-A\|$, $\|A''-A\| \le \varepsilon$, any two compact subsets $V',V'' \subset T$ satisfying $\rho[T_0,V']$, $\rho[T_0,V''] \le \sigma_0$ and $\rho[V',V]$, $\rho[V'',V] \le \sigma_1$, any $V_0' = \{v_1',\dots,v_\ell'\} \in \Sigma^\ell(A',V')$ and any $V_0'' = \{v_1'',\dots,v_\ell''\} \in \Sigma^\ell(A'',V'')$ the following inequality holds

$$|\alpha(A',V')-\alpha(A'',V'')| \le M \max\{\|A'(T')-A''(T'')\|,\|A'(V_0')-A''(S'')\|,\|A'(S')-A''(V_0'')\|\}$$

for any $T' = \{t_1',\dots,t_q'\} \subset V'$, $T'' = \{t_1'',\dots,t_q''\} \subset V''$, $S' = \{s_1',\dots,s_\ell'\} \subset V'$ and $S'' = \{s_1'',\dots,s_\ell''\} \subset V''$ with

$$\rho(t_i,t_i') = \rho[t_i,V'] \le \rho[T_0,V'] \; ; \; \rho(t_i,t_i'') = \rho[t_i,V''] \le \rho[T_0,V''], \; i=1,\dots,q$$

$$\rho(v_j',s_j'') = \rho[v_j',V''] \le \rho[V_0',V''] \; ; \; \rho(v_j'',s_j') = \rho[v_j'',V'] \le \rho[V_0'',V'], \; j=1,\dots,\ell.$$

In the case where $\mathring{\mathcal{C}}(A,T) \cap K$ is non-void the sets V' and V'' are not required to satisfy $\rho[V',V]$, $\rho[V'',V] \le \sigma_1$ (i.e., $\sigma_1 = \infty$).

Proof. Since the subset $V \subset T$ is supposed to be compact, assumption (a) implies the existence of $y^* \in \mathbb{R}^m$ and $\eta > 0$ such that for any $v \in V$,

$$Q(v)y^* \geq 1 + \eta \text{ and } R(v)y^* \geq r(v) + \eta. \tag{15}$$

On the other hand, since the columns of A are equicontinuous on T, there is $\sigma_1 > 0$ such that

$$[s,t \in T \text{ and } \rho(s,t) \leq \sigma_1] \Rightarrow \|A(s)-A(t)\| \leq \min \{\frac{\eta}{3}, \frac{\eta}{3\|y^*\|}\}. \tag{16}$$

Hence, for any $t \in T$ with $\rho[t,V] \leq \sigma_1$ there is $v \in V$ such that $\rho(t,v) \leq \sigma_1$, which, together with (15) and (16), implies

$$Q(t)y^* = Q(v)y^* + (Q(t)-Q(v))y^* \geq 1 + \eta - \|Q(t)-Q(v)\|\|y^*\| \geq 1 + {}^{2\eta}/_3$$

$$R(t)y^* = R(v)y^* + (R(t)-R(v))y^* \geq r(v) + \eta - \|R(t)-R(v)\|\|y^*\| \geq$$

$$r(t) + \eta - |r(t)-r(v)| - {}^{\eta}/_3 \geq r(t) + {}^{\eta}/_3 .$$

Clearly, denoting $\bar{T} \triangleq \{t \mid \rho[t,V] \leq \sigma_1\}$, we deduce that $C(A,\bar{T}) \cap K$ is non-void and (by assumption (a)) is bounded.

Let $T_0 = \{t_1,\dots,t_q\} \in \Delta(A,V)$. Obviously, $T_0 \in \Delta(A,\bar{T})$; let $\varepsilon_0, \sigma_0, M > 0$, as given by Proposition 2 in which $T \triangleq \bar{T}$, A is our given operator, T_0 is the above-mentioned finite set and $\ell \triangleq n + \dim K+1$.

Let $A',A'' \in A(T)$ satisfying $\|A'-A\|, \|A''-A\| \leq \varepsilon_0$ and let two compact subsets $V',V'' \subset T$ satisfying $\rho[T_0,V'],\rho[T_0,V''] \leq \sigma_0$ and $\rho[V',V],\rho[V'',V] \leq \sigma_1$. Thus we have $V',V'' \subset \bar{T}$ and

$$\|A'-A\|_{\bar{T}} \leq \|A'-A\| \leq \varepsilon_0 \ ; \ \|A''-A\|_{\bar{T}} \leq \|A''-A\| \leq \varepsilon_0 .$$

We now prove the required inequality for the operators A' and A" and the sets T_0,V' and V". Let $V_0' = \{v_1',\dots,v_\ell'\} \in \Sigma^\ell(A',V')$ and $V_0'' = \{v_1'',\dots,v_\ell''\} \in \Sigma^\ell(A'',V'')$ where $\ell \triangleq n + \dim K+1$, and choose $S' \triangleq \{s_1',\dots,s_\ell'\} \subset V'$ and $S'' \triangleq \{s_1'',\dots,s_\ell''\} \subset V''$ such that

$$\rho(v_j'',s_j') = \rho[v_j'',V'] \le \rho[V_o'',V'] \qquad j=1,\dots,\ell$$

$$\rho(v_j',s_j'') = \rho[v_j',V''] \le \rho[V_o',V''] \qquad j=1,\dots,\ell.$$

Also choose $T_1' = \{t_1',\dots,t_q'\} \subset V'$ and $T'' = \{t_1'',\dots,t_q''\} \subset V''$ such that

$$\rho(t_i,t_i') = \rho[t_i,V'] \le \rho[T_o,V'] \qquad i=1,\dots,q$$

$$\rho(t_i,t_i'') = \rho[t_i,V''] \le \rho[T_o,V''] \qquad i=1,\dots,q.$$

By Proposition 2 we have

$$|\alpha(A',T' \cup V_o')-\alpha(A'',T'' \cup S'')| \le M \max\{\|A'(T')-A''(T'')\|,\|A'(V_o')-A''(S'')\|\} \triangleq M_1 \tag{17}$$

$$|\alpha(A',T' \cup S')-\alpha(A'',T'' \cup V_o'')| \le M \max\{\|A'(T')-A''(T'')\|,\|A'(S')-A''(V_o'')\|\} \triangleq M_2,$$

which, together with our choice of V_o' and V_o'' and Proposition 1, implies

$$\alpha(A',V') = \alpha(A',T' \cup V_o') \ge \alpha(A',T' \cup S') \ge \alpha(A'',T'' \cup V_o'') - M_2 = \alpha(A'',V'') - M_2$$

$$\alpha(A'',V'') = \alpha(A'',T'' \cup V_o'') \ge \alpha(A'',T'' \cup S'') \ge \alpha(A',T' \cup V_o') - M_1 = \alpha(A',V') - M_1.$$

Thus,

$$|\alpha(A',V') - \alpha(A'',V'')| \le \max\{M_1,M_2\},$$

which, together with (17), completes the proof. □

Note that the condition $\rho[V',V]$, $\rho[V'',V] \le \sigma_1$ is required only to ensure that $V',V'' \subset \bar{T}$. In the case where $\mathring{\mathcal{C}}(A,T) \cap K \neq \phi$, T plays the role of $\bar{T}$, and thus the above condition is no longer required.

Theorem 3. Let T be a compact metric space with metric ρ, $K \subset \mathbb{R}^m$ be a closed convex set, $V \subset T$ be a compact subset and let $A = (f,r,P,Q,R) \in \mathcal{A}(T)$. Suppose that (a) $\mathring{\mathcal{C}}(A,V) \cap K$ is non-void and bounded; and (b) rank $P|_V = n$. Denote $\ell \triangleq n + \dim K+1$. Then,

(i) for any $T_o = \{t_1,\dots,t_q\} \in \Delta(A,V)$ there are $\varepsilon,\sigma_o,\sigma_1$, $M > 0$ such that for any $A' \in \mathcal{A}(T)$ satisfying $\|A'-A\| \le \varepsilon$, any compact subset $V' \subset T$ satisfying

$\rho[T_0,V'] \le \sigma_0$ and $\rho[V',V] \le \sigma_1$, any $V_0 = \{v_1,\dots,v_\ell\} \in \Sigma^\ell(A,V)$ and any $V_0' = \{v_1',\dots,v_\ell'\} \in \Sigma^\ell(A',V')$ the following inequalities hold:

$$|\alpha(A',V')-\alpha(A,V)| \le M \max\{\|A'(T')-A(T_0)\|,\|A'(V_0')-A(S)\|,\|A'(S')-A(V_0)\|\}$$

$$\le M\|A'-A\| + M \max\{\|A(T')-A(T_0)\|,\|A(V_0')-A(S)\|,\|A(S')-A(V_0)\|\}$$

for any $T' = \{t_1',\dots,t_q'\} \subset V'$, $S = \{s_1,\dots,s_\ell\} \subset V$ and $S'=\{s_1',\dots,s_\ell'\} \subset V'$ with

$$\rho(t_i,t_i') = \rho[t_i,V'] \le \rho[T_0,V'] \qquad i=1,\dots,q$$

$$\rho(v_j',s_j) = \rho[v_j',V] \le \rho[V_0',V] \qquad j=1,\dots,\ell$$

$$\rho(v_j,s_j') = \rho[v_j,V'] \le \rho[V_0,V'] \qquad j=1,\dots,\ell;$$

(ii) if, in addition, K is polyhedral or $\mathbb{R}^m$, then for any $T_0 = \{t_1,\dots,t_q\} \in \Delta(A,V)$ and any $V_0 = \{v_1,\dots,v_\ell\} \in \Sigma^\ell(A,V)$ there are ε,σ, $M > 0$ such that for any $A' \in \mathcal{A}(T)$ satisfying $\|A'-A\| \le \varepsilon$ and any compact subset $V' \subset T$ satisfying $\rho[T_0,V'] \le \sigma$ the following inequalities hold:

$$d[\mathcal{S}(A',V'),\mathcal{S}(A,T_0 \cup V_0)] \le$$

$$M|\alpha(A',V')-\alpha(A,V)| + M \max\{\|A'(T')-A(T_0)\|,\|A'(S')-A(V_0)\|\} \le$$

$$M|\alpha(A',V')-\alpha(A,V)| + M\|A'-A\| + M \max\{\|A(T')-A(T_0)\|,\|A(S')-A(V_0)\|\}$$

for any $T' = \{t_1',\dots,t_q'\} \subset V'$ and $S' = \{s_1',\dots,s_\ell'\} \subset V'$ with

$$\rho(t_i,t_i') = \rho[t_i,V'] \le \rho[T_0,V'] \qquad i=1,\dots,q$$

$$\rho(v_j,s_j') = \rho[v_j,V'] \le \rho[V_0,V'] \qquad j=1,\dots,\ell.$$

Proof. (i) The first inequality is a particular case of Theorem 2 where $A''=A$ and $V''=V$; the rest of the sets are $T''=T_0$, $V_0''=V_0$, $S''=S$. For the second inequality we remark that

$$\|A'(T')-A(T_0)\| \le \|A'(T')-A(T')\| + \|A(T')-A(T_0)\| \le \|A'-A\| + \|A(T')-A(T_0)\| \tag{18}$$

Similarly,

$$\|A'(V_0')-A(S)\| \le \|A'-A\| + \|A(V_0')-A(S)\|$$

$$\|A'(S')-A(V_0)\| \le \|A'-A\| + \|A(S')-A(V_0)\|. \tag{19}$$

(ii) Let $T_0 = \{t_1,\ldots,t_q\} \in \Delta(A,V)$ and $V_0 = \{v_1,\ldots,v_\ell\} \in \Sigma^\ell(A,V)$. We have $\alpha(A,V) = \alpha(A,T_0 \cup V_0)$, and $\mathcal{S}(A,T_0 \cup V_0)$ is exactly the set determined by the following inequalities:

$$\begin{aligned}&\pm[F_0Q(T_0,V_0)y - P(T_0,V_0)x] \le \alpha(A,V)Q(T_0,V_0)y\\ &Q(T_0,V_0)y \ge 1,\ R(T_0,V_0)y \ge r(T_0,V_0),\ y \in \mathcal{K}\end{aligned} \tag{20}$$

$$(F_0 = \operatorname{diag} f(T_0,V_0)).$$

On the other hand, for any $A' = (f',r',P',Q',R') \in \mathcal{A}(T)$, any compact subset $V' \subset T$, any $T' = \{t_1',\ldots,t_q'\} \subset V'$ and any $S' = \{s_1',\ldots,s_\ell'\} \subset V'$, the solution set $\mathcal{S}(A',V')$ is included in the set determined by the following inequalities:

$$\begin{aligned}&\pm[F'Q'(T',S')y - P'(T',S')x] \le \alpha(A',V')Q'(T',S')y\\ &Q'(T',S')y \ge 1,\ R'(T',S')y \ge r'(T',S'),\ y \in \mathcal{K}\end{aligned} \tag{21}$$

$$(F' \triangleq \operatorname{diag} f'(T',S'))$$

For any $A' = (f',r',P',Q',R') \in \mathcal{A}(T)$ with $\|A'-A\| \le 1$ we have $\alpha(A',V') \le \|f'\| \le \|f\| + 1$, and thus the set determined by inequalities (21) is included in

$$\bigcup_{y \in \mathcal{C}(A',T' \cup S') \cap \mathcal{K},} \{x \mid \|P(T',S')x\| \le 2(\|f\|+1)(\|Q\|+1)\|y\|\} \times [\mathcal{C}(A',T' \cap S') \cap \mathcal{K}]$$

which by Proposition 2 is uniformly bounded on the parameter set

$$W \triangleq \left\{(A',T',S') \;\middle|\; \begin{aligned}&A' = (f',r',P',Q',R') \in \mathcal{A}(T),\ \|A'-A\| \le \varepsilon\\ &T' = \{t_1',\ldots,t_q'\} \subset T,\ \rho(t_i,t_i') \le \sigma,\ i=1,\ldots,q\\ &S' = \{s_1',\ldots,s_\ell'\} \subset T\end{aligned}\right\}$$

with small enough $\varepsilon,\sigma > 0$.

Thus, the sets determined by (21) are uniformly bounded on $\mathcal{W}$. Now, comparing (20) and (21) and applying Hoffman's theorem (see ref [2]), we obtain a M' > 0 (depending only on $A(T_0,V_0)$) such that for any $(A',T',S') \in \mathcal{W}$:

$$d[\mathcal{S}(A',V'),\mathcal{S}(A,T_0 \cup V_0)] \leq M'\max\{\|F'Q'(T',S') - F_0Q(T_0,V_0)\|,$$
$$\|\alpha(A',V')Q'(T',S')-\alpha(A,V)Q(T_0,V_0)\|,\|A'(T',S')-A(T_0,V_0\|\}$$

After inequality manipulations we find M > 0 such that

$$d[\mathcal{S}(A',V'),\mathcal{S}(A,T_0 \cup V_0)] \leq M|\alpha(A',V')-\alpha(A,V)| + M\|A'(T',S')-A(T_0,V_0)\|$$
$$= M|\alpha(A',V')-\alpha(A,V)| + M\max\{\|A'(T')-A(T_0)\|,\|A'(S')-A(V_0)\|\},$$

which together with (18) and (19), completes the proof. □

In the following corollary we suppose that the columns of A are Lipschitz functions over T. However, as in the previous theorems, the columns of the perturbation A' are only supposed to be continuous.

Corollary 2. Let T be a compact metric space with metric ρ, $K \subset \mathbb{R}^m$ be a closed convex set, $V \subset T$ be a compact and let $A = (f,r,P,Q,R) \in \mathcal{A}(T)$ whose columns are Lipschitz functions. Suppose that (a) $\mathring{\mathcal{C}}(A,V) \cap K$ is non-void and bounded; and (b) rank $P|_V = n$. Denote $\ell = n + \dim K+1$. Then

(i) for any $T_0 \in \Delta(A,V)$ there are $\varepsilon,\sigma,M > 0$ such that for any $A' = (f',r',P',Q',R') \in \mathcal{A}(T)$ satisfying $\|A'-A\| \leq \varepsilon$ and any compact subset $V \subset T$ satisfying $\omega[V',V] \leq \sigma$, the following inequality holds

$$|\alpha(A',V')-\alpha(A,V)| \leq M\|A'-A\| + M\omega[V',V];$$

(ii) if, in addition, K is polyhedral or $\mathbb{R}^m$, then for any $T_0 \in \Delta(A,V)$ and any $V_0 \in \Sigma^{\ell}(A,V)$ there are $\varepsilon,\sigma,M > 0$ such that for any $A' = (R',r',P',Q',R') \in \mathcal{A}(T)$ satisfying $\|A'-A\| \leq \varepsilon$ and any compact subset $V' \subset T$ satisfying $\omega[V',V] \leq \sigma$, the following inequality holds

$$d[\mathcal{S}(A',V'),\mathcal{S}(A,T_0 \cup V_0)] \leq M\|A'-A\| + M\omega[V',V].$$

Proof. By the definitions of $\Delta(\cdot,\cdot), \Sigma^{\ell}(\cdot,\cdot)$ and ω (see introduction), $T_0 \in \Delta(A,V)$, $V_0 \in \Sigma^{\ell}(A,V)$ and $V_0' \in \Sigma^{\ell}(A',V')$ imply $\rho[T_0,V'] \le \rho[V,V'] \le \omega[V',V]$, $\rho[V_0,V'] \le \rho[V,V'] \le \omega[V',V]$ and $\rho[V_0',V] \le \rho[V',V] \le \omega[V',V]$ respectively. Moreover, since the columns of A are Lipschitz functions, there is $L > 0$ such that for any $S = \{s_1,\ldots,s_p\} \subset T$ and $S' = \{s_1',\ldots,s_p'\} \subset T$,

$$\|A(S) - A(S')\| \le L \max_{i=1,\ldots,p} \{\rho(s_i,s_i')\}$$

which, together with Theorem 3, completes the proof. □

We conclude this paper by showing that Theorem 3(ii) implies the continuity of $d[\mathcal{S}(\cdot,\cdot),\mathcal{S}(A,V)]$ at (A,V). Indeed, let $T_0 \in \Delta(A,V)$ and $V_0 \in \Sigma^{\ell}(A,V)$ ($\ell \triangleq n + \dim K+1$), and let (A^k,V^k) be a sequence where $A^k \in \mathcal{A}(T)$ with $\lim_{k\to\infty} \|A^k-A\| = 0$, while $V^k \subset T$ are compact subsets satisfying $\lim_{k\to\infty} \omega[V^k,V] = 0$. Since V is a compact subset of a metric space, we can find a sequence of finite subsets $S^p \subset V$, $p=1,2,\ldots$ such that $cl \bigcup_{p=1}^{\infty} S^p = V$. Clearly, $T^p \triangleq T_0 \cup S^p \in \Delta(A,V)$, $p=1,2,\ldots$. Use now A^k,V^k and T^p instead of A',V',T_0 in Theorem 3(ii), while V_0 is kept fixed; we deduce that for each fixed $p=1,2,\ldots,$

$$\lim_{k\to\infty} d[\mathcal{S}(A^k,V^k),\mathcal{S}(A,T^p \cup V_0)] = 0.$$

Remarking that $\mathcal{S}(A,T^p \cup V_0)$ is bounded (since $T^p \in \Delta(A,V)$), it follows that

$$\lim_{k\to\infty} d[\mathcal{S}(A^k,V^k),\mathcal{S}] = 0 \text{ where } \mathcal{S} \triangleq \bigcap_{p=1}^{\infty} \mathcal{S}(A,T^p \cup V_0). \tag{22}$$

Finally, the continuity of the columns of A and the fact that $V = cl \bigcup_{p=1}^{\infty} T^p \cup V_0$ imply that

$$\mathcal{S} \triangleq \bigcap_{p=1}^{\infty} \mathcal{S}(A,T^p \cup V_0) = \mathcal{S}(A,V),$$

which, together with (22), proves our assertion. □

REFERENCES

[1] Cheney, E.W.: Introduction to approximation theory. McGraw-Hill 1966.

[2] Daniel, J.W.: On perturbations in systems of linear inequalities, *SIAM J. Numer. Anal.*, 10, 299-307 (1973).

[3] Dunham, C.G.: Dependence of best rational Chebyshev approximation on the domain, *Canad. Math. Bull.*, 20, 451-454 (1977).

[4] Dunham, C.B." Continuous dependence in rational Chebyshev approxima= tion, *Journal of Approximation Theory,* 35, 127-131 (1982).

[5] Dunham, C.B.: Chebyshev approximation by rationals with constrained denominators. *Journal of Approximation Theory*, 37, 5-11, (1983).

[6] Flachs, J.: Saddle-point theorems for rational approximation. To appear in *Journal of Approximation Theory.*

[7] Hogan, W.W.: Point-to-set maps in mathematical programming, *SIAM Review,* 15, 591-603 (1973).

[8] Kaufman, E.H. and Taylor, G.D.: Uniform approximation by rational functions having restricted denominators, *Journal of Approximation Theory*, 32, 9-26 (1981).

[9] Krabs, W.: On discretization in generalized rational approximation, *Abh. Math. Sem. Univ. Hamburg* 39, 231-244 (1973).

[10] Robinson, S.M.: Stability theory for systems of inequalities. Part I : Linear systems, *SIAM J. Numer. Anal.,* 12, 754-769 (1975).

[11] Rockafellar, R.T.: *Convex Analysis*, Princeton University Press, 1972.

International Series of
Numerical Mathematics, Vol. 72

The Historical Development of Parametric Programming

Tomas Gal

Fernuniversität Hagen, FRG

The well known tremendous and explosive development of Linear Programming and Operations Research (LP and OR for short, resp.) in general in the 50ties has been connected with the names like CHARNESS, COOPER, GALE, HOFFMAN, KANTOROWICZ, KUHN, TUCKER, WOLFE, to mention only a few of the coryphaei.

G. B. DANZTIG presented 1982 a talk about these times at the Mathematical Programming Conference in Bonn[18]. However, and this was disapointing, he did not mention linear parametric programming by a single word.

I believe that parametric programming (PP for short) does not deserve such kind of handling. This Conference, the Conferences organized by Prof. Fiacco in Washington every year, sessions in big Conferences, today already about 800 publications specialized in the topics, not to speak about publications, in which PP is used in some way as a "by-product", about 13 monographs, sensitivity analysis in all branches of mathematical programming and not only in mathematical programming, methods using parametrization to solve complicated problems of mathematical programming etc., etc., these all are arguments enough that PP cannot be overlooked any more.

Before we start with the historical development of PP, let us first formulate the well known LP-problem: Maximize (or minimize)

$$z = \sum_{j=1}^{n} c_j x_j$$

(LP) s.t.

$$\sum_{j=1}^{n} a_{ij} x_j = b_i, \; i = 1, \ldots, m,$$

$$x_j \geqq 0, \; j = 1, \ldots, n.$$

Starting inquiries into the historical development of a scientific field, the first question which is mostly put is to find out who was the first who posed the problem or who found first some results. As we shall see, in the case of PP this question is not so easy to answer.

Looking in any of lecture books on OR (or on LP) in which there is a chapter on sensitivity analysis or PP or even in the most recent monograph[9] on PP or in the first publication on the history of PP[35], MANNE[72] is mentioned as the first who dealt with the parametrization of the Right Hand Sides (RHS) of (LP). Due to inquiries of Saul I. GASS[38] this is not correct (see also[36]). As GASS says: "This work has been erroneously credited to MANNE due to MANNE's December 1953 report[72] ... . MANNE notes that he is reporting on work by DANTZIG, ORCHARD-HAYS and MARKOWITZ. ORCHARD-HAYS' work was reported in[84] ... . He cites this paper in his book[85] (p. 345) and notes that 'this code and all its successors provided for parametrized RHS's".

Let me note that ORCHAD-HAYS tackled this problem in his unpublished Master thesis in 1952. What he did was the following: Introduce a parameter $\lambda \in \mathbb{R}$ into (LP) in the form

$$(RHS_\lambda) \qquad \sum_{j=1}^{n} a_{ij} x_j = b_i + \lambda b_i', \; i = 1, \ldots, m.$$

Hereby for $\lambda = 0$ the original LP is considered. Hence, it was in fact a postoptimal analysis. Regions (intervals) R_s for λ have been determined such that for any $\lambda \in R_s$, $s = 1, 2, \ldots,$ the problem (LP) with respect to (RHS_λ) has an optimal solution. The reason for introducing λ - at that time - was not a postoptimal analysis but a refinement of the simplex code (see also DANTZIG[19]).

In connection with some applications of LP in practice, MANNE[73] refined the method of ORCHARD-HAYS. His goal was not, however, any refinement of the simplex-code but he concentrated on a parametric analysis as such.

Perhaps because in MANNE's work the term "parametric" appeared the first time, he has become known as the first dealing with PP. There were however others too who dealt with the basic ideas of PP. The corresponding papers were published in 1954[48, 51]. Saul I. GASS writes to this question[38]: "As best as I can make out from my files, we must have started on PP (we called it that from the start, I believe) in 1952. I have a report dated February 1953. ... In about 1952 (!) Walter (JACOBS) posed the basic problem described in[51]" (together with A. J. HOFFMAN). As it is seen there were several people dealing with PP independently of each other in 1952.

Let us have a concise look at what the above mentioned authors did. HOFFMAN and JACOBS[51] investigate a linear model for production planing. Let $r_1, \ldots, r_n$ be given positive constants (shipping requirements in the various months), $x_1, \ldots, x_n$ be nonnegative variables, representing production in the various months, $R_t = \sum_{i=1}^{t} r_i$, $t = 1, \ldots, n$, $X_t = \sum_{i=1}^{t} x_i$, $t = 1, \ldots, n$. Then the stipulation that the shipping requirements to be fulfilled is as $X_t \geqq R_t$, $t = 1, \ldots, n$. Assume that whatever part of the total production at the end of the t-th month that has not been shipped is stored; denote this amount s_t. Then $X_t - R_t = s_t$, $t = 1, \ldots, n$. Define: if $a \in \mathbb{R}$ let $a_+ = \frac{1}{2}(a + |a|)$, i.e. $a_+ = a$ if $a \geqq 0$, $a_+ = 0$ if $a < 0$. With this notation the cost of increasing the production from one month to the other is

$$(x_t - x_{t-1})_+, \quad t = 1, \ldots, n,$$

where $x_o = 0$. The problem is then

(HJ_λ) minimize $\sum_t s_t + \lambda\sum_t (x_t - x_{t-1})_+$

for all nonnegative variables $x_1, \ldots, x_n, s, \ldots, s_n$, where $1/\lambda$ units of increased production cost the same as one unit of storage; the problem is to be solved for all $\lambda \geqq 0$. The authors deal then with 1) the description of some interesting properties of the solution to the problem, and 2) working out a formula for the solution in the special case that the r_t are increasing ($0 < r_1 < r_2 < \ldots < r_n$, i.e. R_t is a convex function of t). Note that in this case the minimizing function becomes $\sum s_i + \lambda x_n$.

Denote K_t the convex envelope of the function R_t (with $R_t = 0$). The authors prove that for any prescribed value of λ, every solution $(X_1, \ldots, X_n)$ of (HJ_λ) satisfies $R_t \leqq X_t \leqq K_t$, and for any value of λ, every solution X satisfies $X_t = R_t$ for each t such that $R_t = K_t$. Writing $f(X) = \sum X_i + \lambda x_n$, the authors prove also the following results:
If $\lambda \leqq \frac{1}{2}(n-k)(n-k+1)$, then $f(X^{(k)}) \leqq f(X^{(k-1)})$.
If $\lambda \geqq \frac{1}{2}(n-k)(n-k+1)$, then $f(X^{(k)}) \geqq f(X^{(k-1)})$, and if $\frac{1}{2}(n-k-1)(n-k) \leqq \lambda \leqq \frac{1}{2}(n-k)(n-k+1)$, then $X^{(k)}$ solves the problem with minimizing $\sum s_i + \lambda x_n$.

GASS and SAATY formulate in the first part[93] of their work the PP with respect to the objective function: Minimize

$$(OF_\lambda) \qquad \sum_{j=1}^{n} (c_j + c_j')x_j,$$

under the conditions given in (LP). In (OF_λ) the c_j's and the c_j''s are constants, $\lambda \in \mathbb{R}$ a parameter. They discuss then some of the basic properties of (OF_λ).

In part 2 [39] they generalize (OF_λ) to an n-parameter case (using here, as I believe, the first time the term multi-parameter), however specialized for the 2-parametric case:

$$(OF_{\lambda_{1,2}}) \qquad \min \sum_{j=1}^{n} (a_j + \lambda_1 b_j + \lambda_2 c_j)x_j$$

s.t. the constraints in (LP).

Two approaches to solve $(OF_{\lambda_{1,2}})$ are described: The double description method and the two-dimensional graph of the corresponding inequalities.

The 3rd part[40] presents a computational algorithm for (OF_λ) which is based on the simplex algorithm assuming nondegeneracy. The procedure starts with a fixed value $\lambda = \delta, -\infty < \delta < \lambda < \phi < \infty$. Then the interval $[\delta, \phi]$ is systematically divided into intervals R_s, s = 1, 2, ..., , today called critical intervals or critical regions[21,34] such that for all $\lambda \in R_s$ (OF_λ) has a finite optimal solution with respect to an optimal basis over a demain defined by $\bigcup_s R_s \cap [\delta, \phi]$.

It is interesting to note that the 3rd part[40] is motivated by the possibility of the existence of more than one objective function to an LP. Looking at the history of multi-criteria decision making the 50ties also witnessed an enormous activity in this field (see e.g. [37] or [104] pp. 58 - 61).

Summarizing, the question "who was first" cannot be answered uniquely. It can just be statet that GASS, JACOBS, ORCHARD-HAYS and SAATY were those who were first starting with analysing LPP-problems already in 1952.

A little bit later I. HELLER[48] has been the first - as far as S. I. GASS and me were able to trace back - who used the term sensitivity analysis. HELLER has in fact used that term for what we call today differential sensitivity analysis.

HELLER starts with a dual pair of LP's, reminds that for feasible x and y

$$b^t y \leq y^T A x \leq c^T x$$

and

$$b^T y = y^T A x = c^T x = \lambda,$$

where λ is the common optimal value of the primal and the dual objective functions. If A, b, c are varied, λ appears as a function $\lambda(A, b, c)$ defined on a certain domain D. He shows that

$$\frac{\delta\lambda}{\delta a_{mv}} = -y_m x_v, \; a_{mv} \text{ the } (m, v)\text{th element of } A,$$

and λ satisfies the system

$$\frac{\delta\lambda}{\delta a_{ij}} = - \frac{\delta\lambda}{\delta b_i} \frac{\delta\lambda}{\delta c_j} \text{ and } \frac{\delta\lambda}{\delta b_i} = y_i, \; \frac{\delta\lambda}{\delta c_j} = x_j,$$

which are known expressions used and analysed again in[21].

The further development of PP lead to terms such as:

- postoptimal sensitivity analysis
- differential sensitivity analysis
- parametric sensitivity analysis
- stability analysis
- perturbation analysis.

During the years the following viewpoints on what is sensitivity analysis crystallized:

1. Investigate the effect of small changes of some or of all of the initial data on the optimal solution (perturbation analysis),
2. Define some specific properties of the optimal solution (in LP e.g. the optimal basis). Investigate then how far one can go in changing some initial data without effecting the required properties (in LP: without changing the optimal basis).

Postoptimal sensitivity analysis includes also the investigation to determine the whole range of parameter(s) for which the given problem has an optimal solution. This kind of analysis is viewed as parametric programming.

The second "peak" in PP research showed up in the second half of the sixtieth (see a corresponding graphical representation in[35]). In these times special cases in linear PP and/or PP in nonlinear programming as well as in other branches of OR where analysed. The first work on a systematic solution procedure for several parameters (multi-parameter) in the RHS or in the objective function coefficients appeared in 1967 [30]. Independently SOKOLOVA [98] proposed another approach in 1968.

In this period also the first monograph on PP has been published[21]. It should be noted that at the same time another monograph has been independently written (see 30) which appeared for several reasons not till 1973 [32].

The peak of the second half of the sixtieth continues into the seventieth. During this period the most methodological and theoretical work has been done by a group at the Humboldt University in East Berlin under the leadership of Professor NOZICKA (this work is summarized in[9]).

They define a parametric mathematical programming problem as to

(P_λ) $\quad \min \{f(x, \lambda) \mid x \in M(\lambda)\}, \ \lambda \in \Lambda,$

$M(\lambda) \subseteq X$, X and Λ metric spaces,

$f : X \times \Lambda \to \mathbb{R} \cup \{-\infty, \infty\}$

with the task: Investigate the properties of the restrictions set mapping

$$M : \lambda \to M(\lambda)$$

the optimal value function

$$\varphi : \lambda \to \varphi(\lambda) := \inf_{x \in M(\lambda)} f(x, \lambda),$$

the optimal set mapping

$$\Psi = \lambda \to \Psi(\lambda) := [x \in M(\lambda) \mid f(x, \lambda) = \varphi(\lambda)\}.$$

For the linear case the parametric programming problem is defined as follows[21, 32, 34]:

(z_λ) $\quad \max z(\lambda) = [c(\lambda)]^T x$

s.t.

$$A(\lambda)x = b(\lambda), \ x \geqq 0, \ \lambda \in \Lambda.$$

The task: Determine a region K (or: $K^* = K \cap \Lambda$) such that for all $\lambda \in K$ (or: $\lambda \in K^*$) (z_λ) has a finite optimal solution, express and investigate the properties of the optimal value function $z_{max}(\lambda)$ over K (or: over K^*), of the relations $x_i(\lambda)$, and $y_{oj}(\lambda)$ over K (or: over K^*), where

$$z_{max}(\lambda) = [c_B(\lambda)]^T x_B(\lambda), \ x_B(\lambda) = B^{-1}(\lambda) b(\lambda),$$

$$y_{oj}(\lambda) = [c_B(\lambda)]^T B^{-1}(\lambda) a^j(\lambda) - c_j(\lambda), \text{ for all } j.$$

Let us note that (z_λ) can be viewed as a special class of (P_λ). Another terminological note: If $\lambda \in \mathbb{R}$ then (P_λ) or (z_λ) is called scalar parametric problem, if $\lambda \in \mathbb{R}^s$, $\lambda = (\lambda_1, \ldots, \lambda_s)^T$ then (P_λ) or (z_λ) is called multiparametric or vectorparametric problem.

As was already mentioned above, we can today look back to a steadily growing research activity (theory, methods, applications) in the field of PP. As a proof for this development the following monographs can be considered:

W. DINKELBACH: Sensitivitätsanalysen und parametrische Programmierung, Springer 1969.[21]

H. NOLTEMEIER: Sensitivitätsanalyse bei diskreten linearen Optimierungsproblemen, Springer 1970.[79]

T. GAL: Betriebliche Entscheidungsprobleme, Sensitivitätsanalyse und parametrische Programmierung, W. de Gruyter 1973 [32]

G. LORENZEN: Parametrische Optimierung und einige Anwendungen, Oldenbourg 1974. [70]

F. NOZICKA, et al.: Theorie der linearen parametrischen Optimierung, Akademie Verlag 1974.[81]

K. B. TLEGENOV et al.: Metody matematiceskogo programmirovanija, Nauka 1975.[101]

U. KAUSSMANN,et al.: Lineare parametrische Optimierung, Akademie Verlag 1976.[57]

T. GAL: Postoptimal analyses, parametric programming, and related topics, McGraw Hill 1979. [34]

K. LOMMATZSCH, ed.: Anwendungen der linearen parametrischen Optimierung, Birkäuser 1979.[69]

R. M. NAUSS: Integer parametric programming, University of Missouri Press 1979.[78]

B. BANK, et al.: Nonlinear parametric optimization. Akademie 1982.[9]

B. BROSOWSKI: Parametric semi-infinite optimization, Peter Lang 1982.[16]

A. L. DONTCHEV: Perturbations, approximations and sensitivity analysis of optimal control systems, Springer 1983.[22]

Let me finally give an overview on selected literature connected with the mainstreams of the development of PP:

Regarding (z_λ) following special cases have been investigated:

$c(\lambda)$ or $A(\lambda)$ or $b(\lambda)$ are analytical functions[28], polynomials[23, 46, 89], linear functions of the parameter (here there are many references, see e.g. [31, 32, 34]).

Theory of (P_λ)[12, 13, 14, 20, 24, 27, 47, 59, 63].

Duality and parametric programming[76, 92]

Sensitivity in nonlinear programming[2, 25, 65, 67]

Generalized parametric equations[90, 91]

Differential stability[15, 26, 41, 42, 43, 66, 68]

Duality gap problems analysed via perturbation of the objective function[62, 99]

Theory of linear parametric programming[80, 81]

Parametrization of the following problems have been investigated:

- complementarity problems[17, 55, 56, 86]
- control of dynamic systems[58]
- fractional programming problems[94]
- geometric programming problems[88]
- integer optimization[10, 44, 60, 74, 79, 100]

- quadratic optimization[64, 75]
- transportation problems [5-8, 29]

Another possibility to use parametric programming is the introduction of parameters into some given mathematical programming problem and the solution of such a problem via parametrization:

- decomposition[1]
- linear vectormaximum problems[33, 103]
- nonconcave problems[50]
- approximation of local solutions of nonlinear optimization problems[2, 4, 27]

Parametric programming has been also applied in practice:

- in the pipeline industry[52, 73]
- for return maximization[96]
- for capital budgeting[53]
- for farm decision problems[30, 45, 49, 71, 87, 97]
- for water ressource problems[61, 83, 102]
- for lot-size problems[11, 54, 77, 82]
- in the meat industry (sausages)[95]

A comprehensive bibliography is to be found in[9, 34]. Let me finish with the statement that looking at the presented very concise historical overview of the development of parametric programming and sensitivity analysis which started in the very early fifties, G. B. DANTZIG should have mentioned PP at least by a single word.

References

1) Abadie, J.M. and Williams, A.C. (1963) Dual and parametric methods in decomposition. In: R.L. Graves, P. Wolfe, eds., Recent advances in mathematical programming (McGraw Hill, New York), 149-158.

2) Armacost, R.L. and Fiacco, A.V. (1975) Second order parametric sensitivity analysis in nonlinear programming and estimates by penalty function methods. Technical Paper No T-324, (The George Washington University, Washington D.C.).

3) Armacost, R.L. and Fiacco, A.V. (1976) Nonlinear programming sensitivity for R.R.S. perturbation: A brief survey and recent second-order-extensions. Technical Paper Series T-334, (The George Washington University, Washington D.C.).

4) Armacost, R.L. and Fiacco, A.V. (1974) Computational experiences in sensitivity analysis for nonlinear programming. Mathematical Programming 6, 301-326.

5) Balachandran, V. (1975) An operator theory of parametric programming for the generalized transportation problem. Part 2: RIM, cost and bound operators. Naval Research Logistics Quarterly 22, 101-125.

6) Balachandran, V. (1975) An operator theory of parametric programming for the generalized transportation problem. Part 3: Weight operators. Naval Research Logistics Quarterly 22, 297-315.

7) Balachandran, V. (1975) An operator theory of parametric programming for the generalized transportation problem. Part 4: Global operators. Naval Research Logistics Quarterly 22, 317-339.

8) Balachandran, V. and Thompson, G.L. (1975) An operator theory of parametric programming for the generalized transportation problem. Part 1: Basic Theory. Naval Research Logistics Quarterly 22, 79-100.

9) Bank, B., Guddat, J., Klatte, D., Kummer, B. and Tammer, K. (1982) Nonlinear parametric optimization (Akademie Verlag, Berlin).

10) Blair, C. and Jeroslow, R.G. (1980) The value function of an integer programm. Working Paper No. MS-80-12 (Georgia Institut of Technology).

11) Bloech, J. (1966) Zum Problem der nachträglichen Änderung industrieller Produktionsprogramme. Zeitschrift für Betriebswirtschaft 36, 186-197.

12) Brosowski, B. (1979) Zur parametrischen linearen Optimierung. II. Eine hinreichende Bedingung für die Unterhalbstetigkeit. Operations Research Verfahren 31, 137-141.

13) Brosowski, B. (1980) On parametric linear optimization. III. A necessary condition for lower semicontinuity. Method of Operations Research 36, 21-30.

14) Brosowski, B. (1981) Parametric semiinfinite optimization (Peter D. Lang Verlag, Frankfurt/Main).

15) Brosowski, B. and Schnatz K. (1980) Parametrische Optimierung, differenzierbare Parameterfunktionen. Zeitschrift für angewandte Mathematik und Mechanik 60, T338-T339.

16) Brosowski, B. (1982) Parametric semi-infinite optimization. In: Brosowski, B. and Martensen, E., eds., Methoden und Verfahren der mathematischen Physik (Verlag Peter D. Lang/ Frankfurt/Main).

17) Cottle, R.W. (1972) Monotone solutions of the parametric linear complementarity problem. Mathematical Programming 3, 210f.

18) Dantzig, G.B. (1983) Reminiscences about the origin of linear programming. In: Mathematical programming - State of the art. Bachem, A., Groetschel, M. and Korte, B., eds., Proceedings of the 11. International Symposium on Mathematical Programming, Bonn 23. - 27. August 1982 (Springer Verlag) 78-86.

19) Dantzig, G.B. (1983) Linear programming and extensions. (Princeton University Press, Princeton, New Jersey).

20) Dantzig, G.B., Folkman, J. and Shapiro, N. (1967) On the continuity of the minimum set of a continuous function. Journal of Mathematical Analysis and Application 17, 519-548.

21) Dinkelbach, W. (1969) Sensitivitätsanalysen und parametrische Programmierung (Springer Verlag, Berlin-Heidelberg-New York).

22) Dontchev, A.L. (1983) Perturbations, Approximations and Sensitivity analysis of optimal control systems. Balakrishnan, A.V. and Thoma, M., eds., Lecture Notes in Control and Information Sciences (Springer Verlag).

23) Dragan, I. (1966) Un algorithme pour la resolution de certain problemes parametriques, avec un seul parametre contenue dans la fonction economique. Revue de Roumain Mathematique pure et Application 11, 4, 447-451.

24) Evans, J.P. and Gould, F.J. (1970) Stability in nonlinear programming. Operations Research 18, 1, 107-118.

25) Fiacco, A.V. (1976) Sensitivity analysis for nonlinear programming using penalty methods. Mathematical Programming 10, 287-311.

26) Fiacco, A.V. (1980) Nonlinear programming sensitivity analysis results using strong second order assumptions. Numerical Optimization of Dynamic Systems, 327-348.

27) Fiacco, A.V. and McCormick, G.P. (1968) Nonlinear programming: Sequential unconstrained minimization techniques (John Wiley).

28) Finkelstein, B.B. (1965) Generalization of the parametric linear programming problem (in Russian). Ekonomicko Matematiceskie Metody 1, 3, 443-450.

29) Fong, C. O. and Rao, M.R. (1975) Parametric studies in transportation-type problems. Naval Research Logistics Quarterly 22, 335-364.

30) Gal, T. (1967) Multiparametric linear programms as an aid for solving farm decision problems (in Czech). (Dissertation for C. Sc., Vysoka Skola Zemedelska, Praha).

31) Gal, T. (1967) Contribution to linear systems programming. Investigation of changes of the a_{ij}-elements of the A-matrix of a linear programming problem. Ekon. Mat. Obzor 3, 446-456.

32) Gal, T. (1973) Betriebliche Entscheidungsprobleme, Sensitivitätsanalyse und parametrische Programmierung (de Gruyter).

33) Gal, T. (1977) A general method for determining the set of all efficient solutions to a linear vectormaximum problem. European Journal for Operational Research 1, 307-322.

34) Gal, T. (1979) Postoptimal analysis, parametric programming and related topics (McGraw Hill, New York).

35) Gal, T. (1980) A 'Historiogramme' of parametric programming. Journal of the Operational Research Society 31, 449-451.

36) Gal, T. (1983) Letter on "A 'Historiogramme' of parametric programming". Journal of the Operational Research Society 34, No 2, 162-163.

37) Gal, T. (1983) On efficient sets in vector maximum problems- A brief survey. In: Essays and Surveys on multiple criteria decision making (Proceedings, MCDM 1982) - Hansen, P., ed., Lecture Notes in Economics and mathematical systems. Beckmann, M. and Krelle, W., eds., (Springer Verlag) 94-114.

38) Gass, S.I. (1983) Private communication.

39) Gass, S.I. and Saaty, T.L. (1955) The parametric objective function 2. Operations Research 3, 395-401.

40) Gass, S.I. and Saaty, T.L. (1955) The computational algorithm for the parametric objective function. Naval Research Logistics Quarterly 2, 39 - 45.

41) Gauvin, J. and Tolle, J.W. (1977) Differential stability in nonlinear programming. SIAM Journal of Control and Optimization 15, 294-311.

42) Geoffrion, A.M. (1966) Strictly parametric programming. I. Basic Theory. Management Science 13, 244-253.

43) Geoffrion, A.M. (1967) Strictly parametric programming. II. Additional theory and computational considerations. Management Science 13, 359-370.

44) Geoffrion, A.M. and Nauss, R. (1977) Parametric and post-optimality analysis in integer linear programming. Management Science 13, 453-466.

45) Glen, J.J. (1980) A parametric programming method for beef cottle ratic formulation. Journal of the Operations Research Society 31, 689-698.

46) Graves, R.L. (1963) Parametric linear programming. In: Graves, R.L. and Wolfe, P., eds., Recend advances in mathematical programming (McGraw Hill, New York) 201-210.

47) Greenberg, H.J. and Pierskalla, W.P. (1972) Extension of the Evans-Gould stability theorems for mathematical programming. Operations Research 20, 143-153.

48) Heller, I. (1954) Sensitivity analysis in linear programming. L.R.P. Seminar. Logistics Research Project (The George Washington University).

49) Hildebrand, P.E. (1959) Farm organization and resource fixity-modification of the LP-model. Ph.D. Thesis (Michigan State University).

50) Hocking, R.R. and Sheppard, R.L. (1971) Parametric solution of a class of nonconvex programs. Operations Research 19, 1742-1747.

51) Hoffman, H.J. and Jacobs, W. (1954) Smooth patterns of production. Management Science 1, 86-91.

52) House, W. (1966) Sensitivity analysis - a case study of the pipeline industry. The Engineering Economist 12, 155-166.

53) House, W. (1967) Use of sensitivity analysis in capital budgeting. Management Services 4, 37-40.

54) Jones, C.H. (1967) Parametric production planning. Management Science 19, 843-866.

55) Jong-Shi Pang (1980) A parametric linear complementarity techniques for optimal portfolio selection with a risk-free asset. Operations Research 28, 927-941.

56) Kaneko, I.K. (1978) A maximization problem related to parametric linear complementarity. SIAM Journal of Control and Optimization 16, 41-55.

57) Kausmann, U., Lommatzsch, K. and Nozicka, F. (1976) Lineare parametrische Optimierung (Akademie Verlag, Berlin).

58) Khalil, K.H. and Kokotovic, P.U. (1979) D-Stability and multi-parameter singular perturbation. SIAM Journal of Control and Optimization 17, 56-65.

59) Klatte, D. (1979) On the lower semicontinuity of optimal sets in convex parametric optimization. Mathematical Programming Study 10, 104-109.

60) Klein, D. and Holm, S. (1979) Integer programming post-optimal analysis with cutting planes. Management Science 25, 64-72.

61) Knowles, W.T., Gupta, I. and Zia Hasan, M. (1976) Decomposition of water distribution networks. AIIE Transactions 8, 443-448.

62) Kortanek, K.O. and Soyster, A.L., On equating the difference between the optimal and marginal values of general convex programms. Journal of Optimization Theory and Applications, to appear, loc. cit.[117]

63) Krabs, W. (1972) Zur stetigen Abhängigkeit des Extremalwertes eines konvexen Optimierungsproblems von einer stetigen Änderung des Problems. Zeitschrift für angewandte Mathematik und Mechanik 52, 359-368.

64) Kummer, B. (1977) Globale Stabilität in der quadratischen Optimierung. Wissenschaftliche Zeitschrift der Humboldt-Universität Berlin, mathematisch-naturwissenschaftliche Reihe 26, Heft 5.

65) Kurcyusz, S. and Zowe, S. (1979) Regularity and stability for the mathematical programming problem in Banach-spaces. Journal for applied Mathematics and Optimization 5, 49-62.

66) Lempio, F. and Maurer, H. (1980) Differential stability in infinite-dimensional nonlinear programming. Journal of Applied Mathematics and Optimization 6, 139-152.

67) Levitin, E.S. (1975) On the local perturbation theory of a problem of mathematical programming in Banach-space. Sowiet Mathematical Doklady Akademii Nauk 16, 1954.

68) Levitin, E.S. (1967) Differentiability with respect to a parameter of the optimal value in parametric problems of mathematical programming. Kibernetika, 44-59.

69) Lommatzsch, K., ed., (1979) Anwendungen der linearen parametrischen Optimierung (Birkhäuser Verlag, Basel und Stuttgart).

70) Lorenzen, G. (1974) Parametrische Optimierung und einige Anwendungen (R. Oldenbourg Verlag, München-Wien).

71) Lyons, D.F. and Dodd, V.A. (1975) The mix-feed problem. In: Haley, K.B., ed., Proceedings of the 7th IFORS International Conference of Operations Research (North-Holland, Amsterdam) 1-15.

72) Manne, A.S. (1953) Notes on parametric linear programming. RAND Corporation Report No p-468.

73) Manne, A.S. (1956) Scheduling of petroleum refinery operations (Harward University Press, Cambridge, Massachusetts).

74) Marsten, R.E. and Morin, T.L. (1977) Parametric integer programming: The right-hand-side-case. Annales of Discrete Mathematics 1, 375-390.

75) McBride, R.D. and Yormark, J.S. (1980) Finding all solutions of a class of parametric quadratic integer programming problems. Management Science 26, 784-795.

76) van Moeseke, P. and Tintner, G. (1964) Base duality theorem for stochastic and parametric programming. Unternehmensforschung 8, 75-79.

77) Müller-Merbach, H. (1962) Sensibilitätsanalyse der Losgrößenbestimmung. Unternehmensforschung 6, 79-88.

78) Nauss, R.M. (1979) Integer parametric programming. (University of Missouri Press).

79) Noltemeier, N. (1970) Sensitivitätsanalyse bei diskreten linearen Optimierungsproblemen. Lecture Notes in Operations Research and Mathematical Systems (Springer Verlag Berlin).

80) Nozicka, F. (1972) Über eine Klasse von linearen einparametrischen Optimierungsproblemen. Mathematik, Operationsforschung und Statistik 3, 159-194.

81) Nozicka, F., Guddat, J., Hollatz, H. and Bank, B. (1974) Theorie der linearen parametrischen Optimierung. (Akademie-Verlag, Berlin).

82) Oberhoff, W.-D. (1975) Integrierte Produktionsplanung. Deterministische Entscheidungsmodelle zur Planung optimaler Losgrößen und des optimalen Produktionsprogramms. (Studienverlag Dr. N. Brockmeyer, Bochum).

83) O'Laoghaire, D.T. and Himmelblau, D.M. (1972) Modelling and sensitivity analysis for planning decision in water resource expanses. Water Research Bulletin 8, 653-668.

84) Orchard-Hays, W. (1955) Notes on linear programming (Part 6): The RAND code for the simplex method (SX4). Rand Corporation, Rep. 10, 1440.

85) Orchard-Hays, W. (1968) Advanced linear programming computing techniques(McGraw Hill, New York).

86) Panne van de, C. (1975) Methods for linear and a quadratic programming (North-Holland, Amsterdam).

87) Panne van de, C. and Popp, J. (1968) Minimum cost cattle feed under probalistic protein constraints. Management Science 9, 405-430.

88) Peterson, E.L. (1977) The duality between suboptimization and parameter deletion with application to parametric programming and decomposition theory in geometric programming. Mathematics of Operations Research 2, 311-319.

89) Ritter, K. (1963) Über Probleme parameterabhängiger Planungsrechnung. DVL Bericht Nr. 238 (Porz-Wahn).

90) Robinson, S.M. (1975) Stability theory for systems of inequalities. Part I: Linear system. SIAM Journal of Nummerical Analysis 12, 754-769.

91) Robinson, S.M. (1976) Stability theory for systems of inequalities. Part II. Differentiable nonlinear systems. SIAM Journal of Nummerical Analysis 13, 497-513.

92) Rockafellar, R.T. (1967) Duality and stability in extremum problems involving convex functions. Pacific Journal of Mathematics 21, 167-187.

93) Saaty, T.L. and Gass, S.I. (1954) The parametric objective function. 1. Operations Research 2, 316-319.

94) Saxena, P. C. and Aggarwal, S. P. (1980) Parametric linear fractional functional programming. Economical Computings and Economical Research 87-97.

95) Schimitzek, P. (1981) Rezepturoptimierung für Fleischerzeugnisse. Dissertation, (RWTH Aachen, Inst. f. Wirtschaftswissenschaften).

96) Seelbach, H. (1968) Rentabilitätsmaximierung bei variablem Eigenkapital. Zeitschrift für Betriebswirtschaft 8, 237-256.

97) Sengupta, J.K. and Sanyal, B.C. (1970) Sensitivity analysis methods for a crop-mix-problem in LP. Unternehmensforschung 14, 2-26.

98) Sokolova (Grygarova), L. (1968) Problem viceparametrickeho programovani. Ekon. Mat. Obzor 4, 44-68.

99) Soyster, A.L. (1981) An objective function perturbation with economic interpretation. Management Science 27, 231-237.

100) Suzuki, H. (1978) A generalized knapsack problem with variable coefficients. Mathematical Programming 15, 162-176.

101) Tlegenov, K.B., Kaltschaief, K.K. and Zapletin, P.P. (1975) Mathematical programming methods (in Russian). (Nauka, Alma-Ata).

102) Wels, G.R. (1975) A sensitivity analysis of simulated river basin planning for capital budgeting decisions. Computings and Operations Research 2, 49-54.

103) Yu, P.L. and Zeleny, M. (1975) The set of all nondominated solutions in linear cases and a multicriteria simplex method. Journal of Mathematical Analysis and Applications 49, 430-468.

104) Zeleny, M. (1982) Multiple criteria decision making. McGraw Hill).

Prof. Dr. Dr. T. Gal, Lehrgebiet für Operations Research und Wirtschaftsmathematik, Fernuniversität, 5800 Hagen, W. Germany

International Series of
Numerical Mathematics, Vol. 72

OPTIMIZATION PROBLEMS ON EXTREMAL ALGEBRAS: NECESSARY AND SUFFICIENT CONDITIONS FOR OPTIMAL POINTS

Siegfried Helbig

Johann Wolfgang Goethe-Universität
Fachbereich Mathematik
Robert Mayer-Straße 6-1o
Frankfurt/Main
West-Germany

1. Introduction

When we consider optimization problems with convex restriction functions and a convex objective function, it is well-known that the set of feasible points is convex and each local optimal point is global. These properties are also fulfilled, if convexity is replaced by several generalizations (quasi-convexity, pseudo-convexity, pointwise-convexity (only the second property)).

We replace the usual concept of convexity by the so-called extremally convexity. For this we have to define a general algebraic structure over a set F and a topology on F - the so-called topological extremal algebra (see ZIMMERMANN (3), (4) and ZIMMERMANN & JUHNKE (5)). When we define extremally convex functions in this sense, the above properties are fulfilled in this class of problems (see (3)). In this article, we introduce (after the presentation of the extremal algebra, some examples and some topological aspects) the semi-infinite-extremally minimization problems EP(∇) and ELP(∇).

In chapter 4., we give a necessary and sufficient condition for minimal points of ELP(∇) (the extremally-linear problem). An Example shows that this condition is not sufficient for minimal points of EP(∇) (the general problem). This is in contrast to semi-infinite optimization problems (see BROSOWSKI (1)).

For applications see for example CUNNINGHAME-GREEN (2).

2. The axioms of an extremal algebra, extremally convexity and topological aspects

Definition 1.: (see (3), (4), (5)) Suppose that two binary operations $\oplus$ and $\bullet$ are given on a set F. This set F is called extremal algebra, when

(A1) F is a commutative semigroup with respect to $\oplus$ and $\bullet$.

(A2) $x \bullet (y \oplus z) = x \bullet y \oplus x \bullet z$ for all x, y, z $\in$ F (distributivity)

(A3) There exist two different elements $\bar{0}$ and $\bar{1} \in F$ such that:
$x \bullet \bar{0} = \bar{0}$, $x \bullet \bar{1} = x$, $x \oplus \bar{0} = x$ for all $x \in F$

(A4) $x \oplus y$ = either x or y for all x, y $\in$ F

(A5) $x \leq y \Leftrightarrow x \oplus y = y$

Properties 2.: (see (3)) Suppose arbitrary x, y $\in$ F.

1.) '$\leq$' is an ordering relation on F

2.) $x \geq 0$ for all $x \in F$

3.) $x \leq y$, $z \in F$ $\Rightarrow$ $x \oplus z \leq y \oplus z$

4.) $x \leq y$, $z \in F$ $\Rightarrow$ $x \bullet z \leq y \bullet z$

In the sequel, we need two further axioms (see also (3)).

(A6) $(x, y, z \in F \text{ with } x < y \leq z) \Rightarrow (\exists\ \alpha \in F$ with $\overline{0} < \alpha < \overline{1}$ such that: $x < \alpha \bullet z < y)$

(A7) $(x \in F \text{ with } x \neq \overline{0}) \Rightarrow (\exists\ x^{-1} \in F$ such that: $x \bullet x^{-1} = \overline{1}$.

Example 3.: (see also (2), (3), (4) and (5))

a) $F := \{x \in \mathbb{R} \mid x \geq 0\}$

Let $x \oplus y := \max(x,y)$ and $x \bullet y := xy$ (usual multiplication

Then the following statements can be proved:

(i) $\overline{0} = 0$ and $\overline{1} = 1$;

(ii) the relation '$\leq$' is the usual 'smaller or equal'-relation on the real numbers.

b) $F := \mathbb{R} \cup \{-\infty\}$

Let $x \oplus y := \max(x,y)$ and $x \bullet y := x + y$.

Then the following statements can be proved:

(i) $\overline{0} = -\infty$ and $\overline{1} = 0$;

(ii) the relation '$\leq$' is the usual 'smaller or equal'-relation on the real line.

c) $F := \{x \in \mathbb{R} \mid x > 0\} \cup \{\infty\}$

Let $x \oplus y := \min(x,y)$ and

$$x \bullet y := \begin{cases} xy & \text{(usual multiplication) if } x,y \neq \infty \\ \infty & \text{if either } x \text{ or } y \text{ is} \end{cases}$$

Then the following statements can be proved:

(i) $\overline{0} = \infty$ and $\overline{1} = 1$;

(ii) the relation '$\leq$' is the usual 'greater or equal'-relation on the real line.

d) $F := \{x \in \mathbb{R} \mid a \leq x \leq b \text{ with } a, b \in \mathbb{R} \text{ and } a < b\}$

Let $x \oplus y := \max(x,y)$ and $x \bullet y := \min(x,y)$.

Then the following statements can be proved:

(i) $\overline{0} = a$ and $\overline{1} = b$

(ii) the relation '$\leq$' is the usual 'smaller or equal'-relation on the real numbers.

All sets F considered above are extremal algebras, but only the sets from examples a), b) and c) fulfille the axioms (A6) and (A7) because for the extremal algebra in d),

(A7) is not satisfied for all $x \neq \overline{1}$.

Let F be a set satisfieing (A1) - (A7) and extend the operations $\oplus$ and $\bullet$ on $F^n := F \times F \times F \ldots \times F$ where $n \in N$. For elements $x = (x_1, \ldots, x_n)$ and $y = (y_1, \ldots, y_n) \in F^n$ and $\alpha \in F$, we define:

(i) 'extremal sum': $x \oplus y := (x_1 \oplus y_1, \ldots, x_n \oplus y_n) \in F^n$

(ii) 'extremal scalar multiplication': $\alpha \bullet x := (\alpha \bullet x_1, \ldots, \alpha \bullet x_n)$

(iii) 'extremal inner product':

$$x \bullet y := (x,y) := x_1 \bullet y_1 \oplus \ldots \oplus x_n \bullet y_n = \sum_{i=1}^{n\,\oplus} x_i \bullet y_i \in F$$

In consequence of (A4) there exists at least one $j \in 1, \ldots, n$ such that:

$$(x,y) = \sum_{i=1}^{n\,\oplus} x_i \bullet y_i = x_j \bullet y_j .$$

<u>Definition 4.:</u> A set $A \in F^n$ is called <u>extremally-convex</u> (<u>e-convex</u>), when for all $x, y \in A$ and for all $\alpha, \beta \in F$ with $\alpha \oplus \beta = \overline{1}$ the following implication holds:

$$x, y \in A \quad \Rightarrow \quad \alpha \bullet x \oplus \beta \bullet y \in A.$$

Define the sets U_{ab}, $U_{\overline{0}b}$ and U_a with $a, b \in F$ by:

(i) $U_{ab} := \{ z \in F \mid a < z < b \}$

(ii) $U_{\overline{0}b} := \{ z \in F \mid \overline{0} \leq z < b \}$

(iii) $U_a := \{ z \in F \mid a < z \}$

Suppose an arbitrary $y \in F$; then U_{ab} (resp. $U_{\overline{0}b}$ or U_a) is an element of $\mathfrak{U}(y)$, if and only if $y \in U_{ab}$ (resp. $U_{\overline{0}b}$ or U_a).

<u>Theorem 5.:</u> a) The sets U_{ab}, $U_{\overline{0}b}$ and U_a are extremally-convex for arbitrary $a, b \in F$.

b) The family of sets $(\mathfrak{U}(y))_{y \in F}$ forms the basis of neighbourhoods of a topology Υ on F.

c) Υ is hausdorffsch.

<u>Proof:</u> a), b) trivial.

c) Suppose $x, y \in F$ with $x < y$. According to (A6) there exists an element $\alpha \in F$ with $\overline{0} < \alpha < \overline{1}$ such that

$$x < \alpha \bullet y < y$$

because of $x < y \leq y$. Define neighbourhoods U_{ab} and U_{cd} by:

$a := \bar{0}$, $b := \alpha \cdot y$, $c := \alpha \cdot y$, $d := \alpha^{-1} \cdot y$.

Since $a \leq x < \alpha \cdot y = b$ and $c = \alpha \cdot y < y < \alpha^{-1} \cdot y$ (since $\bar{1} < \alpha^{-1}$)

i) $x \in U_{ab} \in \mathcal{U}(x)$ and $y \in U_{cd} \in \mathcal{U}(y)$

ii) $U_{ab} \cap U_{cd} = \emptyset$.

So the assertion holds. ∎

F^n is equipped with the product topology Υ^n. All neighbourhoods of a point $y \in F^n$ are contained in the set $\mathcal{U}(y)$. For the next chapters we need the following statements:

Theorem 6.: Suppose $a, b \in F^n$ with $a_i < b_i$ for $i = 1, \ldots, n$ and define

$$Q_{ab} := \{ z \in F^n \mid a_i \leq z_i \leq b_i, \; i = 1, \ldots, n \}.$$

The set Q_{ab} is compact in the topology Υ^n. ∎

Corollary 7.: Every infinite bounded sequence in F^n has a cluster point. ∎

We remark that a function $f : F \to F$ is continous in $x \in F$, if for all $U \in \mathcal{U}(f(x))$ there exists a neighbourhood V of x with $f(V) \subset U$. A function f which is continuous over a compact set $I \subset F$ attains its minimum and maximum in I.

3. The extremally optimization problems

Let $T \subset F$ be a compact set and define the sets $EC(T)$ and $EC(T,F^n)$ by

$$EC(T) := \{ f : T \to F \mid f \text{ is continuous on } T \}$$

$$EC(T,F^n) := \{ g : T \times F^n \to F \mid g \text{ is continuous} \}$$

which are equipped with the topologies

$$\Upsilon^T := \prod_{t \in T} \Upsilon \quad \text{and}$$

$$\Upsilon^{T,n} := \prod_{t \in T} \Upsilon \times \Upsilon^n.$$

For each triplet $\sigma := (A,b,p)$ of continuous mappings $A : T \times F^n \to F$,

$b : T \to F$ and

$p : F^n \to F$, we define the general extremal minimization problem

$$\text{EP}(\sigma) \qquad \text{Minimize } p : F^n \to F$$

$$\text{subject to: } A(t,x) \geqslant b(t) \text{ for all } t \in T.$$

Let $B : T \to F^n$ be a continuous mapping and $p \in F^n$; then consider the corresponding extremally-linear minimization problem

$$\text{ELP}(\sigma) \qquad \text{Minimize } (p,x) = \sum_{i=1}^{n\,\oplus} p_i \cdot x_i$$

$$\text{subject to: } (B(t),x) \geqslant b(t) \text{ for all } t \in T,$$

where $(.,.)$ denotes the 'extremal inner product' which we present in chapter 2.

Without loss of generality, we cann assume that $b(t) > \bar{0}$ for all $t \in T$ because $\bar{0} \leqslant (B(t),x) \in F$ for all $t \in T$ and all $x \in F^n$ (see Property 2.a)), from which we can conclude that those restricitons where $b(t) = \bar{0}$ are redundant.

In the sequel, we only consider the problem ELP(σ); we define the feasible set, extremal value and the set of minimal points as follows:

$$Z_\sigma := \{ x \in F^n \mid (B(t),x) \geqslant b(t) \}$$

$$E_\sigma := \inf_{x \in Z} (p,x)$$

$$P_\sigma := \{ x \in Z_\sigma \mid (p,x) = E_\sigma \}$$

<u>Theorem 8.:</u> a) Z_σ is extremally-convex

b) P_σ is extremally-convex

c) If $\alpha \cdot v \oplus \beta \cdot u \in P_\sigma$ for $\alpha, \beta \in F$ with $\alpha \oplus \beta = \bar{1}$ and $u, v \in Z_\sigma$, then either u or v is an element of P_σ.

Proof: a) Let x, y be elements of Z_∇ and $\alpha, \beta \in F$ with $\alpha \oplus \beta = \bar{1}$; then we have for arbitrary $t \in T$

$$\begin{aligned}(B(t), \alpha\cdot x \oplus \beta\cdot y) &= \alpha\cdot(B(t),x) \oplus \beta\cdot(B(t),y)\\ &\geqslant \alpha\cdot b(t) \oplus \beta\cdot b(t) = (\alpha\oplus\beta)\cdot b(t) = b(t)\end{aligned}$$

according to (A3), the properties 2.b)&c), and the properties of the extremal inner product. For this $\alpha\cdot x \oplus \beta\cdot y \in Z_\nabla$.

b) Let $x, y \in P_\nabla$, i. e. $E_\nabla = (p,x) = (p,y)$, and $\alpha, \beta \in F$ with $\alpha \oplus \beta = \bar{1}$; then we have

$$\begin{aligned}E_\nabla &= (\alpha\oplus\beta)\cdot E_\nabla = \alpha\cdot E_\nabla \oplus \beta\cdot E_\nabla\\ &= \alpha\cdot(p,x) \oplus \beta\cdot(p,y) = (p, \alpha\cdot x \oplus \beta\cdot y);\end{aligned}$$

therefore $\alpha\cdot x \oplus \beta\cdot y \in P_\nabla$.

c) Assume that $u, v \notin P_\nabla$, i. e. $E_\nabla < (p,u)$ and $E_\nabla < (p,v)$; then we have for $\alpha, \beta \in F$ with $\alpha \oplus \beta = \bar{1}$

$$\begin{aligned}E_\nabla &= \alpha\cdot E_\nabla \oplus \beta\cdot E_\nabla < \alpha\cdot(p,u) \oplus \beta\cdot(p,v)\\ &= (p, \alpha\cdot u \oplus \beta\cdot v) = E_\nabla. \quad ■\end{aligned}$$

For the corresponding sets of EP(∇), the statements of theorem 8. are not fulfilled in general (see Example 12.).

There are several types of minimization problems ELP(∇) as is indicated by the following example:

Example 9.: Let F be the extremal algebra from example 3.a), namely $F := \{x \in \mathbb{R} \mid x \geqslant 0\}$, $x \oplus y := \max(x,y)$, $x \cdot y := xy$, and $n = 2$.

a) Suppose $T = \{1,2\}$; $p = (1, 2/3)$; $B(1) = (1, 1/2)$; $B(2) = (1/2, 1)$
 $b(1) = b(2) = 1$.

ELP(∇): $\min\ (p,x) = \max(x_1, \frac{2}{3}x_2)$

subject to

$(B(1),x) = \max(x_1, \frac{1}{2}x_2) \geqslant 1$

$(B(2),x) = \max(\frac{1}{2}x_1, x_2) \geqslant 1$

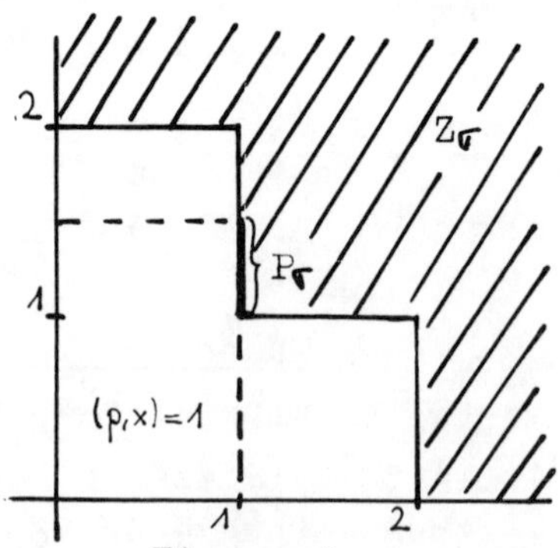

Figure 1.

b) Suppose $T = [1,2]$; $B(t) = (t^{-1}, t)$ and $b(t)$ for all $t \in T$.

$ELP(\sigma)$: $\min\ (p,x)$

subject to: $(B(t),x) = \max(t^{-1} \cdot x_1, tx_2) \geq b(t)\ \forall t \in T$

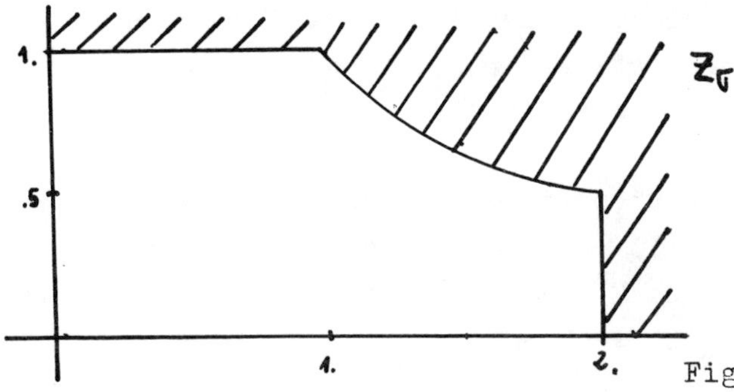

Figure 2.

4. Characterisations of minimal points of $ELP(\sigma)$

Let $t_0 \notin T$ and $T_0 := T \cup \{t_0\}$, which is compact too. Extend the functions $B : T \rightarrow F^n$ and $b : T \rightarrow F$ by setting:

$b(t_0) = \bar{1}$ and

$$(B(t_0),x) := \begin{cases} E_\sigma \cdot (p,x)^{-1}, & \text{if } E_\sigma > \bar{0} \text{ and } (p,x) > \bar{0} \\ \alpha \cdot (p,x)^{-1}, & \text{if } E_\sigma = \bar{0} \text{ and } (p,x) > \bar{0} \text{ and} \\ & \quad \bar{0} < \alpha < (p,x) \quad \text{(A6)} \\ \bar{1}, & \text{if } E_\sigma = \bar{0} \text{ and } (p,x) = \bar{0} \\ \bar{0}, & \text{if } E_\sigma > \bar{0} \text{ and } (p,x) = \bar{0} \end{cases}$$

Since $b(t) > \bar{0}$ for all $t \in T_0$, $b(t)^{-1} \cdot (B(t),x)$ is continuous on T_0 for each $x \in F^n$.

For arbitrary $v \in F^n$ we define the sets

$$M^0_{\sigma,v} := \{ t \in T_0 \mid b(t)^{-1} \cdot (B(t),v) = \min_{s \in T_0} b(s)^{-1} \cdot (B(s),v) \}$$

$$M_{\sigma,v} := \{ t \in T \mid b(t) = (B(t),v) \}$$

Then we have the following sufficient condition for minimal points of $ELP(\sigma)$ which is comparable with a result of semi-infinite optimization (see (1), p. 48; notice that the theorem in (1) was proved for the general non-linear problem):

<u>Theorem 10.:</u> Let $v_o \in F^n$ with the property that $\forall_{v \in F^n} \; \exists_{t \in M^o_{\sigma, v_o}}$ such that $(B(t), v_o) \geq (B(t), v)$.

Then v_o is a minimal point of $ELP(\sigma)$.

<u>Proof:</u> Define for $x \in F^n$ a function f by

$$f_\sigma(x) := \min_{t \in T_o} b(t)^{-1} \cdot (B(t), x).$$

Because $b(t)^{-1} \cdot (B(t), x)$ is continuous on T_o and attains its minimum in a $t \in T_o$, $f_\sigma(x) \in F$ for all $x \in F^n$.

Let $x \in F^n$ be arbitrary; by assumption, there exists a $t \in M^o_{\sigma, v_o}$ such that

$$f_\sigma(v_o) = b(t)^{-1} \cdot (B(t), v_o) \geq b(t)^{-1} \cdot (B(t), x)$$
$$\geq f_\sigma(x)$$

Therefore: $f_\sigma(x) \leq f_\sigma(v_o)$ for all $x \in F^n$ (*)

For $x \notin Z_\sigma$, there exists $t \in T$ with $b(t) > (B(t), x)$ which is equivalent to

$$b(t)^{-1} \cdot (B(t), x) < \overline{1};$$

this implies: $f_\sigma(x) < \overline{1}$.

For $x \in Z_\sigma$, there is $E_\sigma \leq (p, x)$.

<u>Case 1:</u> $E_\sigma > \overline{0}$ and $(p, x) > \overline{0}$

This implies: $b(t_o)^{-1} \cdot (B(t_o), x) = E_\sigma \cdot (p, x)^{-1} \leq \overline{1}$.

<u>Case 2:</u> $E_\sigma = \overline{0}$ and $(p, x) > \overline{0}$

This implies: $b(t_o)^{-1} \cdot (B(t_o), x) = \alpha \cdot (p, x)^{-1} < \overline{1}$.

<u>Case 3:</u> $E_\sigma = \overline{0}$ and $(p, x) = \overline{0}$

This implies: $b(t_o)^{-1} \cdot (B(t_o), x) = \overline{1}$.

<u>Case 4:</u> $E_\sigma > \overline{0}$ and $(p, x) = \overline{0}$

This implies: $(p, x) < E_\sigma \leq (p, x)$. This is a contradiction because $x \in Z_\sigma$.

In all cases, we have: $f_\sigma(x) \leq \overline{1}$. (#)

Now assume that $f_\sigma(v_o) < \overline{1}$.

<u>Cases 1&2:</u> $E_\sigma \geq \overline{0}$ and $(p, v_o) > \overline{0}$

I.) $t_o \notin M^o_{\sigma, v_o}$; let $s > \overline{1}$ be arbitrary; then for all points

t in M^o_{σ, v_o}: $(B(t), v_o) < (B(t), s v_o)$.

Since $s\cdot v_0 \in F^n$, this is a contradiction to the assumption of this theorem.

II.) $t_0 \in M^0_{\sigma,v_0}$; this implies (since $f_\sigma(v_0) < \overline{1}$ by assumption)

$$\overline{0} \leq E_\sigma < (p,v_0).$$

Therefore $P_\sigma \neq \emptyset$. If $P_\sigma = \emptyset$, there does not exist any $x \in Z_\sigma$ with $(p,x) = E_\sigma$, and $E_\sigma = \infty$.

Let $v \in P_\sigma$; we conclude that

$$b(t)^{-1}\cdot(B(t),v_0) < \overline{1} \leq b(t)^{-1}\cdot(B(t),v)$$

for all $t \in M^0_{\sigma,v_0} \setminus \{t_0\}$ and

$$b(t_0)^{-1}\cdot(B(t_0),v_0) < 1 = E_\sigma \cdot (p,v)^{-1}.$$

From this inequality we deduce that

$$(B(t),v_0) < (B(t),v) \qquad \text{for all } t \in M^0_{\sigma,v_0}.$$

This is a contradiction.

<u>Case 3:</u> $(p,v_0) = \overline{0} = E_\sigma$

This implies that $t_0 \notin M^0_{\sigma,v_0}$ because $b(t_0)^{-1}\cdot(B(t_0),v_0) = \overline{1}$.

Let $s > \overline{1}$ be arbitrary and conclude as above (Case 1&2, Part I.))

<u>Case 4:</u> $E_\sigma > \overline{0}$ and $(p,v_0) = \overline{0}$

In this case, we have $f_\sigma(x) \leq f_\sigma(v_0) = \overline{0}$ for all $x \in F^n$ which implies

$$f_\sigma(x) = \min_{t \in T_0} b(t)^{-1}\cdot(B(t),x) = \overline{0} \text{ for all } x \in F^n.$$

Set $y_i := \overline{1}$ for $i = 1, \ldots ,n$; we easily conclude that $b(t)^{-1}\cdot(B(t),y) > \overline{0}$ for all $t \in T$ and $(p,y) > \overline{0}$ because there must exist $i, j \in \{1, \ldots , n\}$ such that $B(t)_i \neq \overline{0}$ and $p_j \neq \overline{0}$ (since $b(t) > \overline{0}$ and $E_\sigma > \overline{0}$).

But for the element $y \in F^n$, we have

$$f_\sigma(y) = \min_{t \in T_0} b(t)^{-1}\cdot(B(t),y) > \overline{0}.$$

Consequently, we have: $f_\sigma(v_0) = \overline{1}$. (✡)

This equation reveals that $v_0 \in Z_\sigma$ because for all $t \in T$ $b(t)^{-1}\cdot(B(t),v_0) \geq \overline{1}$ which is equivalent to

$$(B(t),v_0) \geq b(t) \qquad \text{for all } t \in T.$$

Now we have to show that $v_o \in P_\sigma$.

Case 1: $E_\sigma > \bar{0}$ and $(p,v_o) > \bar{0}$

By (✡), we get: $E_\sigma \cdot (p,v_o)^{-1} \geq \bar{1}$, which is equivalent to $(p,v_o) \leq E_\sigma$.

Case 2: $E_\sigma = \bar{0}$ and $(p,v_o) > \bar{0}$

This implies: $\alpha \cdot (p,v_o)^{-1} = b(t_o)^{-1} \cdot (B(t),v_o) < \bar{1}$; this case cannot occur since (✡).

Case 3: $E_\sigma = \bar{0}$ and $(p,v_o) = \bar{0}$

Obviously.

Case 4: $E_\sigma > \bar{0}$ and $(p,v_o) = \bar{0}$

This case cannot occur because $v_o \in Z_\sigma$.

In the relevant cases, we have $v_o \in P_\sigma$ which concludes the proof. ∎

The background idea of the proof of this theorem is from (1), even though the technique of this proof is quite different.

Similar to semi-infinite optimization problems (see (1)), we have the following conclusion:

Corollary 11.: Let v_o be an element of Z_σ with the property that

$$\forall_{v \in F^n} \ \exists_{t \in M_{\sigma,v_o} \cup \{t_o\}} \text{ such that } (B(t),v_o) \geq (B(t),v).$$

Then v_o is a minimal point of ELP(σ).

Proof: Suppose $t \in M^o_{\sigma,v_o} \setminus \{t_o\}$. Because of $f_\sigma(x) \leq \bar{1}$ for all $x \in F^n$ (see (#)) and $v_o \in Z_\sigma$, we have

$$\bar{1} \leq b(t)^{-1} \cdot (B(t),v_o) \leq \bar{1}.$$

This implies $t \in M_{\sigma,v_o}$ and $M^o_{\sigma,v_o} \subset M_{\sigma,v_o} \cup \{t_o\}$.

Now suppose $t_o \notin M^o_{\sigma,v_o}$. Since $f_\sigma(v_o) \leq \bar{1}$ and $v_o \in Z_\sigma$, we obtain

$$b(t_o)^{-1} \cdot (B(t_o),v_o) > \bar{1},$$

so that the cases 2 and 4 cannot occur. Regard the other two cases.

<u>Case 1:</u> $E_\sigma > \bar{0}$ and $(p,v_0) > \bar{0}$

Therefore $b(t_0)^{-1} \cdot (B(t_0),v_0) = E_\sigma \cdot (p,v_0)^{-1} > \bar{1}$ which implies $E_\sigma > (p,v_0)$.

This is a contradiction to $v_0 \in Z_\sigma$.

<u>Case 3:</u> $E_\sigma = \bar{0}$ and $(p,v_0) = \bar{0}$

This implies $b(t_0)^{-1} \cdot (B(t_0),v_0) = \bar{1}$ which is a contradiction to the above conclusion.

We have still to proof that

$$t \in M_{\sigma,v_0} \quad \text{implies} \quad t \in M^o_{\sigma,v_0}$$

Suppose $t \in M_{\sigma,v_0}$, i. e. $b(t) = (B(t),v_0)$, hence

$$\bar{1} = b(t)^{-1} \cdot (B(t),v_0)$$

and thus

$$\bar{1} < b(s)^{-1} \cdot (B(s),v_0) \quad \text{for all } s \in T \setminus M_{\sigma,v_0}$$

It suffices to proof that $f_\sigma(v_0) = \bar{1}$; then we conclude since $t_0 \in M^o_{\sigma,v_0}$:

$$\bar{1} = b(t)^{-1} \cdot (B(t),v_0) = b(t_0)^{-1} \cdot (B(t_0),v_0) = \min_{s \in T_0} b(s)^{-1} \cdot (B(s),v_0)$$

for all $t \in M_{\sigma,v_0}$.

If we assume $f_\sigma(v_0) < \bar{1}$, we lead this assumption to a contradiciton like in the proof in theorem 10.

Consequently, we have

$$M^o_{\sigma,v_0} = M_{\sigma,v_0} \cup \{t_0\}$$

Now we are able to conclude that $v_0 \in P_\sigma$ (according to theorem 10). ∎

The next example clears up the question wether it is possible to generalize the charcterization theorem and its corollary to the general extremal minimization problem of type EP(σ) or not.

<u>Example 12.:</u> Suppose that F is the extremal algebra from example 3b), namely F := $\mathbb{R} \cup \{-\infty\}$ with the operations $\oplus$:= max and $\bullet$:= + (usual addition). Let A be an elemnt of EC(T,F^n), i. e. A : T $\times F^n \rightarrow$ F is a continuous mapping, and n = 2. Then define for arbitrary v $\in F^n$ the set

$$M_{\sigma,v} := \{ t \in T \mid b(t) = A(t,v) \}.$$

Suppose that T = $\{t_1, t_2\}$, p = (0,0), $b(t_1) = b(t_2) = 1$, and $A(t_1,x) = \sin(x_1 \cdot \frac{\pi}{2}) \oplus \cos(x_2)$, $A(t_2,x) = x_1 \oplus x_2$.

Notice that 0 is the real number zero, but $\overline{0} = -\infty$.

Then we have the following minimization problem:

EP(σ): $\quad \min (p,x) = \min x_1 \oplus x_2$

subject to

$$A(t_1,x) = \sin(x_1 \cdot \tfrac{\pi}{2}) \oplus \cos(x_2) \geq 1$$
$$A(t_2,x) = x_1 \oplus x_2 \geq 1$$

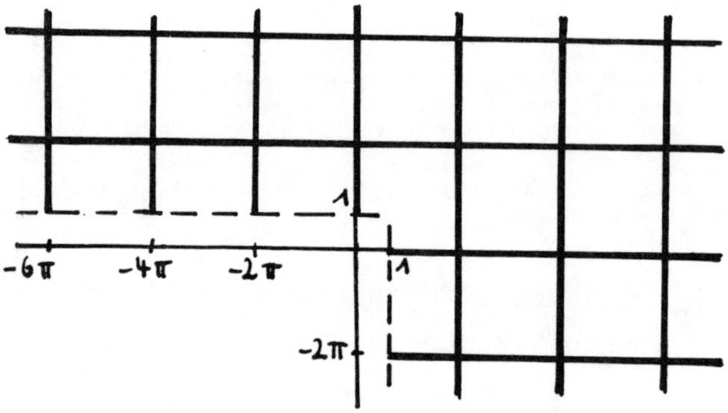

Figure 3.

As is indicated in figure 3.

$$Z_\sigma = \bigcup_{-k\in N_0} \{2k\pi\} \times [1,\infty) \cup \bigcup_{k\in N} \{2k\pi\} \times F \cup$$
$$\bigcup_{-k\in N_0} [1,\infty) \times \{2k\pi\} \cup \bigcup_{k\in N} F \times \{2k\pi\}$$

An element v is a minimal point of EP(σ), i. e. v $\in P_\sigma$, if it has the form v = (1,2kπ) or v = (2kπ,1) with -k $\in N_0$.

Therefore: $E_\sigma = (p,v) = 1 \oplus 2k\pi = 1$ (if $-k \in N_0$)

Suppose we have $v_0 = (2\pi,0)$; then $M_{\sigma,v_0} = \{t_1\}$ and

$$A(t_1,v_0) = 1 > A(t_1,x) \quad \text{for all } x \in F^n,$$

so the assumption of corollary 11. is fulfilled. But for $x = (1,0)$ we have

$$(p,v_0) = 0 \cdot 2\pi = 2\pi > 1 = 0 \cdot 1 = (p,x).$$

Therefore v_0 is not a minimal point although the condition of corollary 11. is satisfied.

We see that the extremally-linearity of the extremal inner product and the extremally-convexity of Z_σ play an important roll. ∎

The next theorem shows that the condition given in theorem 10. is also necessary for a minimal point of ELP(σ).

<u>Theorem 13.:</u> Suppose that v_0 is a minimal point of ELP(σ), i. e. $v_0 \in P_\sigma$. Then

$$\forall_{v \in F^n} \quad \exists_{t \in M_{\sigma,v_0} \cup \{t_0\}} \quad \text{such that } (B(t),v_0) \geqslant (B(t),v)$$

i. e. the above condition is necessary for a minimal point of ELP(σ).

<u>Proof:</u> Let $W_\sigma := \{ v \in F^n \mid (p,v) < (p,v_0) \}$

<u>i)</u> $v \notin W_\sigma$:

This implies $(p,v_0) \leqslant (p,v)$. Again we have to consider the four cases.

<u>Case 1:</u> $E_\sigma > \bar{0}$ and $(p,v) > \bar{0}$

This implies: $(B(t_0),v_0) = E_\sigma \cdot (p,v_0)^{-1}$
$> E_\sigma \cdot (p,v)^{-1} = (B(t_0),v)$.

<u>Case 2:</u> $E_\sigma = \bar{0} = (p,v_0)$ and $(p,v) \geqslant \bar{0}$

This implies: $(B(t_0),v_0) = \bar{1} > \alpha \cdot (p,v)^{-1} = (B(t_0),v)$.

<u>Case 3:</u> $E_\sigma = \bar{0} = (p,v_0) = (p,v)$

This implies: $(B(t_0),v_0) = \bar{1} \geqslant \bar{1} = (B(t_0),v)$.

<u>Case 4:</u> $E_\sigma > \bar{0}$ and $(p,v) = \bar{0}$

Since $v \notin W_\sigma$ the inequality

$$(p,v) = \bar{0} < E_\sigma = (p,v_0) \leq (p,v) = \bar{0}$$

forms a contradiction.

In all relevant cases the assertion is fulfillde if $v \notin W_\sigma$.

ii) $v \in W_\sigma$:

This implies $(p,v) < (p,v_0)$. We assert that $M_{\sigma,v_0} \neq \emptyset$ in this case. If $M_{\sigma,v_0} = \emptyset$, then $E_\sigma = \bar{0}$ which implies $W_\sigma = \emptyset$.

If we assume $E_\sigma > \bar{0}$, we can deduce from $M_{\sigma,v_0} = \emptyset$ that

$$(B(t),v_0) > b(t) \qquad \text{for all } t \in T.$$

Define $s := \max_{t \in T} b(t) \cdot (B(t),v_0)^{-1} < \bar{1}$. The function $b(t) \cdot (B(t),v_0)^{-1}$ is continuous on T since $(B(t),v_0) > \bar{0}$ (since $v_0 \in Z_\sigma$) and attains its maximum on T.

If we define now $w_i := s \cdot v_{0i}$ $(i = 1, \ldots, n)$, we see that

$$b(t) = b(t) \cdot (B(t),v_0)^{-1} \cdot (B(t),v_0) \leq s \cdot (B(t),v_0)$$
$$= (B(t), s \cdot v_0) = (B(t),w)$$

for all $t \in T$.

Since $s < \bar{1}$, we have

$$(p,w) = p_i \cdot w_i = s \cdot p_i \cdot v_{0i} < p_i \cdot v_{0i} \leq (p,v_0) = E_\sigma$$

for at least one $i \in \{1, \ldots, n\}$.

Since $w \in Z_\sigma$, this is a contradiction to $v_0 \in P_\sigma$. So we conclude that $W_\sigma = \emptyset$ if $M_{\sigma,v_0} = \emptyset$.

We have to consider two cases:

Case I: $v_i \leq v_{0i}$ for $i = 1, \ldots, n$

Take an arbitrary $t \in M_{\sigma,v_0}$; the the following inequality holds

$$(B(t),v) = B(t)_i \cdot v_i \leq B(t)_i \cdot v_{0i} \leq (B(t),v_0)$$

for at least one $i \in \{1, \ldots, n\}$.

From this we can conclude the assertion.

Case II: $v_{0j} < v_j$ for at least one $j \in \{1, \ldots, n\}$

Define an element $w \in F^n$ by $w_i := v_i \oplus v_{0i}$ for $i = 1, \ldots, n$; we assert that $w \in P_\sigma$.

1.) Let $t \in T$ be arbitrary; then at least one $i \in \{1, \ldots, n\}$ satisfies the inequality

$$b(t) \leq (B(t),v_o) = B(t)_i \cdot v_{oi} \leq B(t)_i \cdot w_i \leq (B(t),w)$$

2.) At least one $l \in \{1, \dots, n\}$ satifies

$$(p,v_o) = p_l \cdot v_{ol} > (p,v) \geq p_l \cdot v_l,$$

hence $v_l < v_{ol} = v_{ol} \oplus v_l = w_l$ since $p_l \neq \bar{0}$.

For all j with $v_j > v_{oj}$, we have

$$(p,v_o) > (p,v) \geq p_j \cdot v_j = p_j \cdot (v_j \oplus v_{oj}) = p_j \cdot w_j.$$

Subsuming this, we conclude that

$$E_\sigma = (p,v_o) = p_l \cdot v_{ol} = p_l \cdot w_l \geq p_i \cdot w_i$$

with $i \in \{1, \dots, n\}$, $i \neq l$.

From 1.) we deduce $w \in Z_\sigma$; together with 2.), we have $w \in P_\sigma$.
By case I, there exists a $t \in M_{\sigma,w}$ such that

$$(B(t),w) \geq (B(t),v)$$

since $v_i \leq w_i$ for $i = 1, \dots, n$.
Assume that for this $t \in M_{\sigma,w}$ the inequality

$$b(t) = (B(t),w) < (B(t),v_o) = B(t)_j \cdot v_{oj}$$

is valid for at least one $j \in \{1, \dots, n\}$;
this implies $t \notin M_{\sigma,v_o}$.

Then we have the estimation

$$B(t)_j \cdot v_j \leq B(t)_j \cdot w_j \leq (B(t),w) < B(t)_j \cdot v_{oj} = B(t)_j \cdot w_j$$

since $v_j < v_{oj}$ and $B(t)_j \neq \bar{0}$.
On account of this contradiction, we see that $t \in M_{\sigma,v_o}$.
From this we deduce the assertion that

$$(B(t),v_o) = (B(t),w) \geq (B(t),v)$$

for an element $t \in M_{\sigma,v_o}$ if $v \in W_\sigma$. ∎

<u>Definition 14.:</u> Let $Q \subset F^n$ be an arbitrary set.
Then the set

$$ECE(Q) := \left\{ \sum_{i=1}^{r} \alpha_i \cdot q^i \,\middle|\, \alpha_i \in F,\ r \in \mathbb{N},\ q^i \in Q \right\}$$

is called the extremally-convex envelope of Q.

The next theorem gives a further sufficient condition for minimal points of ELP(σ).

<u>Theorem 15.:</u> Let v_0 be an alement of Z_σ. If $p \in ECE(\{B(t) \in F^n \mid t \in M_{\sigma,v_0}\})$,

then $v_0 \in P_\sigma$.

<u>Proof:</u> Because of the assumption, there exists $\rho_t \in F$ for $t \in M_{\sigma,v_0}$ with

$$p = \sum_{t \in M_{\sigma,v_0}}^{\oplus} \rho_t \cdot B(t).$$

Let v be an arbitrary feasible point. The the following estimate holds:

$$(p,v_0) = \sum_{t \in M_{\sigma,v_0}}^{\oplus} \rho_t \cdot (B(t),v_0) = \sum_{t \in M_{\sigma,v_0}}^{\oplus} \rho_t \cdot b(t)$$

$$\leq \sum_{t \in M_{\sigma,v_0}}^{\oplus} \rho_t \cdot (B(t),v) = (p,v).$$

Since v is arbitrary, it follows: $v_0 \in P_\sigma$. ∎

The following example will show that this theorem, which can be compared with the Kuhn-Tucker-theorem in usual linear optimization, is not reversible in general.

<u>Example 16.:</u> Take as extremal algebra the set $F := \{x \in \mathbb{R} \mid x \geq 0\}$ and the operations $\oplus$:= max and $\bullet$:= $\cdot$ (see example 3a)).
Take ELP(σ) from example 9a).
We can take the element $v_0 = (1,3/2)$ as an element from P_σ and we have $M_{\sigma,v_0} = \{1\}$, but

$$p = \begin{pmatrix} 1 \\ 2/3 \end{pmatrix} = \rho_1 \cdot \begin{pmatrix} 1 \\ 1/2 \end{pmatrix} = \rho_1 \cdot B(1)$$

has no solution $\rho_1 \in F$.
We presume that there exists at least one $v \in P_\sigma$ such that the reversal of theorem 16. is valid.
In this example we have to take $v = (1,1) \in P_\sigma$. Then we have

$M_{\sigma, v_0} = \{1,2\}$ and the system of equations

$$p = \begin{pmatrix} 1 \\ 2/3 \end{pmatrix} = \varrho_1 \cdot \begin{pmatrix} 1 \\ 1/2 \end{pmatrix} \oplus \varrho_2 \cdot \begin{pmatrix} 1/2 \\ 1 \end{pmatrix} = \varrho_1 \cdot B(1) \oplus \varrho_2 \cdot B(2)$$

has the solution $\varrho_1 = 1$, $\varrho_2 = 2/3$.
There must be 'enough' points in the set of active constraints.

5. Remarks

In chapter 4., the parameter $\sigma := (B,b,p)$ of our extremal minimization problem was fixed. But we can also consider extremally-linear minimization problems, where the restriction vector b (or resp. the objective function p) is variable. This means that b (resp. p) is the parameter of the optimization problem. In account of this we write ELP(b), Z_b, E_b, and P_b (resp. ELP(p), E_p, and P_p) instead of ELP(σ), Z_σ, E_σ, and P_σ. We are interested in continuouity properties of the mappings

$$Z : L_{Bp} \rightarrow POT(F^n)$$
$$E : L_{Bp} \rightarrow F$$
$$P : L_{Bp} \rightarrow POT(F^n)$$

(resp. $E : L_{Bb} \rightarrow F$ and $P : L_{Bb} \rightarrow POT(F^n)$), where L_{Bp} (resp. L_{Bb}) denotes denotes the parameter space of our poblem, which is defined by

$$L_{Bp} := \{ b \in EC(T) \mid P_b \neq \emptyset \}$$

(resp. $L_{Bb} := \{ p \in F^n \mid P_p \neq \emptyset \}$).
Another question is how to decompose the parameter space.

In further papers, the author will give the results of the above mentioned questions.

6. References

(1) Brosowski, B. (1982): Parametric semi-infinite optimization; Peter Lang Verlag, Frankfurt/Main and Bern.

(2) Cunninghame-Green, R. (1979): Minimax Algebra; Springer-Verlag, Berlin, Heidelberg, New York.

(3) Zimmermann, K. (1979): Extremally convex functions; Wissenschaftliche Zeitschrift der Pädagogischen Hochschule 'N.K.Krupskaja' Halle, XVII, Heft 1, 3 - 7.

(4) Zimmermann, K. (1980): Solution of some optimization problems on the extremal algebra; Mathematical methods Operations Research 1, Akad. Kiado, Budapest, 179 - 185.

(5) Zimmermann, K. & F. Juhnke (1979): Lineare Gleichungs- und Ungleichungssysteme auf extremalen Algebren; Wissenschaftliche Zeitschrift der Technischen Hochschule Otto von Guericke Magedeburg 23, Heft 1, 103 - 106.

Siegfried Helbig, Taläckerstraße 28 a, 6230 Frankfurt/Main 80, West-Germany

International Series of
Numerical Mathematics, Vol. 72

ON THE LOCAL STRUCTURE OF THE FEASIBLE SET IN SEMI-INFINITE OPTIMIZATION

H.Th. Jongen, G. Zwier

Twente University of Technology,
Dept. Applied Mathematics, The Netherlands.

Abstract. We present the generic local structure of the feasible set in semi-infinite optimization, in the case that the index set of the inequality constraints is an interval. In fact, via a decomposition theorem, the feasible set will be described, locally and up to diffeomorphism, by means of a set of elementary functions.

Acknowledgement. We are indebted to Wolfgang Wetterling for the careful reading of the first draft of the present manuscript. Moreover, we would like to thank Peter Kenderov for sending us the new book of V.I. Arnold et al. (Ref. [1]). With the aid of this book we found recent important contributions to this area by soviet mathematicians.

Introduction.

A semi-infinite optimization problem (SIP) is usually formulated in the following way :

$$\text{(SIP)} \quad \text{Minimize } f \text{ on } M, \tag{1}$$

where $f : R^n \to R$ is a continuous function (object function) and $M \subset R^n$ the feasible set, defined by

$$M = \{x \in R^n \,|\, h_i(x) = 0,\ i = 1,..,m,\ g_j(x,y_j) \geq 0 \text{ for all } y_j \in Y_j \text{ and all } j = 1,..,s\}. \tag{2}$$

In (2), $h_i : R^n \to R$, $g_j : R^n \times R^{m_j} \to R$ are continuous functions, whereas $Y_j \subset R^{m_j}$ are compact subsets. The adjective "semi-infinite" is related with the fact that the sets Y_j need not to be discrete point sets. An important class of problems which can be written within the formalism of semi-infinite optimization, is formed by Chebyshev-approximation, i.e. the problem of finding a best approximation of a continuous function - uniformly w.r.t. a given compact set - by means of a given n-parameter family of functions (cf. [9], [11]).

For $\bar{x} \in M$, the points $\bar{y}_j^{(r)}$ for which $g_j(\bar{x},\bar{y}_j^{(r)}) = 0$, are (global) minima for the function $g_j(\bar{x},\cdot)$ restricted to Y_j. Suppose that g_j is of class C^2 and Y_j is a compact C^2-manifold with boundary (i.e. at each point $y_j \in Y_j$, the set Y_j is locally C^2-diffeomorphic to an open neighborhood of the origin in $H^p \times R^q$, H^p being the nonnegative orthant in R^p). Under the additional assumption that $\bar{y}_j^{(r)}$ is a nondegenerate (local) minimum for $g_j(\bar{x},\cdot)$ restricted to Y_j (i.e. we have strict complementarity and nondegeneracy of the corresponding quadratic form), the implicit function theorem yields locally defined functions $y_j^{(r)}(x)$ of class C^1, each $y_j^r(x)$ being a nondegenerate local minimum for $g_j(x,\cdot)$ restricted to Y_j. Moreover, the composite functions $\phi_j^{(r)}(x) := g_j(x,y_j^{(r)}(x))$ are of class C^2 (cf. [5], [7], [8], [11], [14]). In the latter case, the set M is (locally) described by means of a finite number of equality constraints (h_i) and a <u>finite</u> number of inequality constraints ($\phi_j^{(r)}$). In general, however, one cannot avoid that at some points $\bar{x} \in M$ the corresponding points $\bar{y}_j^{(r)} \in Y_j^{(r)}$ are degenerate minima. In fact, some degeneracies remain stable under small C^2-perturbations of the functions g_j.
The aim of this paper is the study of the local structure of the feasible set M for functions h_i, g_j in "general position". We restrict ourselves to the following special case :

$$M = \{x \in R^n \mid F(x,y) \geq 0 \text{ for all } y \in [a,b]\}, \tag{3}$$

where $a,b \in R$, $a < b$. Moreover, we assume that $F : R^n \times R \to R$ is of class C^∞. For "general position"-reasoning, this differentiability assumption is mild, because of the fact that the C^∞-functions are C_S^k-dense in the space of C^k-functions, where C_S^k denotes the strong (Whitney) C^k-topology (cf. [10]). The ideas and results of our paper are easily extended to sets M of the type

(2), in the case that every set Y_j is an interval $[a_j,b_j]$, where $a_j < b_j$ (or $a_j = b_j$).

Statement of the results.

By $C^\infty(R^{n+1},R)$ we denote the space of functions from R^{n+1} to R, which are infinitely many times differentiable. A vector $z \in R^{n+1}$ will always be partitioned as $z = (x,y)$, where $x \in R^n$ and $y \in R$.
For $F \in C^\infty(R^{n+1},R)$, D_xF denotes the row vector of partial derivatives with respect to $x = (x_1,\ldots,x_n)$. In the sequel, a and b are fixed real numbers, and $a < b$.

Definition 1. Let $F \in C^\infty(R^{n+1},R)$ and $z = (x,y)$. For $k \geq 1$, a point $y \in [a,b]$ is called a zero-point for $F(x,\cdot)$ of order k, if the following holds at z :

$$\left(\frac{\partial}{\partial y}\right)^i F = 0, \; i = 0,1,\ldots r, \quad \text{and} \quad \left(\frac{\partial}{\partial y}\right)^{r+1} F \neq 0, \tag{4}$$

where

$$r = k, \text{ resp. } k\text{-}1, \text{ according to } y \in (a,b), \text{ resp. } y = a,b. \tag{5}$$

Definition 2. If $y \in [a,b]$ is a zero-point for $F(x,\cdot)$ of order k, $k \geq 1$, then y is said to be a zero-point of finite order. In that case, we define $\mathrm{Ord}(y) = k$ and introduce the following associated set of vectors S(y) :

$$S(y) = \left\{ \left(\frac{\partial}{\partial y}\right)^i D_xF, \; i = 0,1,\ldots,k\text{-}1\right\}. \tag{6}$$

For $F \in C^\infty(R^{n+1},R)$ we define the minimum function $\phi_F : R^n \to R$,

$$\phi_F(x) = \min_{y \in [a,b]} F(x,y) \tag{7}$$

In this way, the set M in (3) becomes:

$$M = \{x \in R^n \mid \phi_F(x) \geq 0\}. \tag{8}$$

Remark 1. If $\bar{x} \in M$ and $\psi_F(\bar{x}) > 0$, then M contains an open neighborhood of $\bar{x}$. Hence, the interesting case is : $\bar{x} \in M$ and $\psi_F(\bar{x}) = 0$.

Definition 3. In case $\psi_F(x) = 0$, we define the extremal set $E_F(x)$:

$$E_F(x) = \{y \in [a,b] \mid F(x,y)=0\} \tag{9}$$

Remark 2. Note that every $y \in E_F(x)$ is a global minimum for $F(x,\cdot)\big|_{[a,b]}$.

Definition 4. An $F \in C^\infty(R^{n+1},R)$ is said to be regular (w.r.t. [a,b]) if the following conditions hold, whenever $\psi_F(x) = 0$:

R_1. The set $E_F(x)$ is finite, say $\{y_1,\dots,y_p\}$.
R_2. Every $y \in E_F(x)$ is a zero-point for $F(x,\cdot)$ of finite order.
R_3. $\dim.\mathrm{span}\,[\bigcup_{i=1}^{p} S(y_i)] = \sum_{i=1}^{p} \mathrm{Ord}(y_i)$.

Remark 3. Obviously, in Definition 4, R2 implies R1. Moreover, if F is regular and $y \in E_F(x) \cap (a,b)$, then R2 implies that y is a zero-point of odd order.

For $k \geq 1$, we introduce the following functions $\eta_k, \phi_k : R^k \to R$.

$$\eta_k(x) = \min_{y \in [-1,+1]} \{y^{k+1} + \sum_{i=1}^{k} x_i y^{k-i}\} \quad (x \in R^k) \tag{10}$$

$$\phi_k(x) = \min_{y \in [0,1]} \{y^k + \sum_{i=1}^{k} x_i y^{k-i}\} \quad (x \in R^k) \tag{11}$$

Remark 4. From (10), (11), we see that $\eta_1(x) = \phi_1(x) = x$.

Put:

$$\mathcal{F} = \{F \in C^\infty(R^{n+1},R) \mid F \text{ regular (w.r.t. } [a,b])\}. \tag{12}$$

Theorem 1. Let $F \in \mathcal{F}$ and let the set M be defined by (8). Suppose that $\bar{x} \in M$ and $\psi_F(\bar{x}) = 0$. Put $E_F(\bar{x}) = \{\bar{y}_1,\dots,\bar{y}_p\}$, where $\bar{y}_1 < \bar{y}_2 < \dots < \bar{y}_p$.

Define $k_i = \mathrm{Ord}(\bar{y}_i)$, $i = 1,\ldots,p$, and let t_i, $i = 0,1,\ldots,p$, be defined recursively as follows : $t_0 = 0$, $t_i = t_{i-1} + k_i$.
Then, there exist an (R^n-) open neighborhood $\mathcal{Q}$ of $\bar{x}$, an open neighborhood V of $0 \in R^n$, and a C^∞-diffeomorphism $\Phi : \mathcal{Q} \to V$, $u = \Phi(x)$, sending $\bar{x}$ onto 0, such that $\Phi(M \cap \mathcal{Q})$ is defined by means of the following system of inequalities:

$$\psi_{k_i}(u_{t_{i-1}+1}, \ldots, u_{t_i}) \geq 0, \; i = 1,\ldots,p, \tag{13}$$

where

$$\left.\begin{array}{l} \psi_{k_1} = \phi_{k_1} \text{ if } \bar{y}_1 = a, \; \psi_{k_p} = \phi_{k_p} \text{ if } \bar{y}_p = b, \\ \text{and } \phi_{k_i} = \eta_{k_i} \text{ otherwise.} \end{array}\right\} \tag{14}$$

In (14), the functions η_k, ϕ_k, refer to (10), (11). □

By C^k_S we denote the strong (Whitney) C^k-topology for $C^\infty(R^{n+1},R)$ (cf. [10]).

<u>Theorem 2.</u> The set F is C^k_S -dense for all k. Moreover, F is C^k_S-open for all $k \geq r$, where $r = n$, resp. $n+1$, according to n even, resp. n odd. □

<u>Proof of Theorem 1.</u>

<u>Definition 5.</u> Let $x, u \in R^n$, and $y, v \in R$. A local C^∞-coordinate transformation $(x,y) \to (u,v)$, sending the origin onto itself, is said to be of Type I, resp. of Type II, if it has the following structure :

$$\text{Type I : } (u,v) = (\xi(x), \eta(x,y)), \tag{15}$$

$$\text{Type II : } (u,v) = (\xi(x), y\eta(x,y)), \; \eta(0) > 0. \tag{16}$$

<u>Remark 5.</u> If a coordinate transformation is of Type II, then $v = 0$ whenever $y = 0$, and the orientation of the y-axis is preversed.

The proof of Theorem 1 mainly depends on the following lemma.

<u>Lemma 1.</u> Let $F \in C^\infty(R^{n+1},R)$, $z = (x,y)$ and suppose that $\bar{y} \varepsilon [a,b]$ is a zero-point for $F(\bar{x},\cdot)$ of order $k(\geq 1)$. Consider the following two typical cases:

<u>Case I.</u> $\bar{z} = 0$, $\bar{y} \in (a,b)$ and $S(\bar{y})$ linearly independent;

$$\text{put } \varepsilon = \text{sign} \left(\frac{\partial}{\partial y}\right)^{k+1} F(0). \tag{17}$$

<u>Case II.</u> $\bar{z} = 0$, $\bar{y} = a$, and $S(\bar{y})$ linearly independent;

$$\text{put } \varepsilon = \text{sign} \left(\frac{\partial}{\partial y}\right)^{k} F(0). \tag{18}$$

In Case I (resp. Case II) there exists a local C^∞-coordinate transformation Q of Type I (resp. Type II) such that:

$$F \circ Q^{-1}(u,v) = \varepsilon v^{k+1} + \sum_{i=1}^{k} u_i v^{k-i} \quad (\text{resp.} = \varepsilon v^k + \sum_{i=1}^{k} u_i v^{k-i}). \tag{19}$$

<u>Proof.</u> We start with Case I. From (4), (17), it follows that $F(0,y) = \varepsilon y^{k+1} h(y)$, where $h(0) > 0$. Put $(w,q) = (x, yh(y)^{1/k+1})$; shortly, $(w,q) = Q_1(x,y)$. Then, Q_1 is a local coordinate transformation of Type II and hence, of Type I. Put $\tilde{F} = F \circ Q_1^{-1}$ Consequently, $\tilde{F}(0,q) = \varepsilon q^{k+1}$. From singularity theory (cf. [4]) we know that the function $\varepsilon q^{k+1} + \sum_{i=1}^{k-1} \gamma_i q^{k-i}$, where $\gamma_i \in R$, is a universal unfolding for εq^{k+1}. So, locally, there exist C^∞-functions $\alpha_i(w)$, $i = 1,\dots,k$, $\beta(w,q)$, satisfying:

$$\tilde{F} = \varepsilon \beta^{k+1} + \sum_{i=1}^{k} \alpha_i \beta^{k-i}, \tag{20}$$

where $\alpha_i(0) = 0$, $i = 1,\dots,k$, and $\beta(0,q) \equiv q$. (21)

A straightforward calculation, using (20), (21), shows:

$$\begin{aligned} \left(\frac{\partial}{\partial q}\right)^{j} D_w \tilde{F}(0) &= (j!) D\alpha_{k-j}(0), \quad j = 0,1,\dots,k-1, \\ \left(\frac{\partial}{\partial q}\right)^{k} D_w \tilde{F}(0) &= \varepsilon((k+1)!) D_w \beta(0). \end{aligned} \tag{22}$$

Let A_F, resp $A_{\tilde{F}}$ be the $(k+1)\times n$ matrix with $\left(\frac{\partial}{\partial y}\right)^{i-1} D_x F(0)$, resp. $\left(\frac{\partial}{\partial q}\right)^{i-1} D_w \tilde{F}(0)$, as its $i^{\underline{th}}$ row, $i = 1,\dots,k, k+1$. From the special structure of Q_1 we see that

$$A_{\tilde{F}} = B\, A_F, \tag{23}$$

where B is a nonsingular, lower triangular $(k+1)x(k+1)$ matrix. A combination of (6), (17), (22), (23), shows that $\{D\alpha_i(0),\ i = 1,\dots,k\}$ is a linearly independent set. Put

$$u_i = \alpha_i(w),\ i = 1,\dots,k \quad , \ v = \beta(w,q), \tag{24}$$

Next, we choose C^∞-functions $\phi_j(w)$, $j = k+1,\dots,n$, all of them vanishing at the origin and such that $\{D\alpha_i(0), D\phi_j(0), i = 1,\dots,k, j = k+1,\dots,n\}$ is a linearly independent set. Finally, put

$$u_j = \phi_j(w),\ j = k+1,\dots,n, \tag{25}$$

Note that (24), (25) define a local coordinate transformation Q_2 of Type I. Now by the very construction. $F\circ(Q_2\circ Q_1)^{-1}$ takes the form as in (19), Case I. (A moment of reflection shows that the value of ε is irrelevant, in case k is even).

We proceed with the proof in Case II, <u>$k \geq 2$.</u> Firstly, we can follow the reasoning as is done in Case I, including Formula (23), by <u>replacing every entry of k by k-1</u>. Then, instead of (24), we put:

$$\left.\begin{array}{l} u_i = \alpha_i(w),\ i = 1,\dots,k-1,\ u_k = \beta(w,0) \\ \ v = \beta(w,q) - \beta(w,0), \\ u_j = \phi_j(w),\ j = k+1,\dots,n, \end{array}\right\} \tag{26}$$

where $\phi_j(w)$ are C^∞-functions having the property - besides $\phi_j(0) = 0$ - that $\{D\alpha_j(0), j = 1,\dots,k-1, D_w\beta(0), D\phi_j(0), j = k+1,\dots,n\}$ is a linearly independent set.

Note that $\beta(w,q) - \beta(w,0) = q\gamma(w,q)$ where $\gamma(0) = \frac{\partial}{\partial q}\beta(0)$ (= 1 in view of (21)). Consequently, (26) defines a local coordinate transformation Q_2 of Type II, and $F\circ(Q_2\circ Q_1)^{-1}$ takes the following form:

$$\varepsilon(v + u_k)^k + \sum_{i=1}^{k-1} u_i(v+u_k)^{k-i-1}. \tag{27}$$

Next, we put:

$$u_i' = \text{coefficient of } v^{k-i} \text{ in (27)},\ i = 1,\dots,k, \text{ and } v' = v. \tag{28}$$

Together with the functions ϕ_j as in (26), we see that (28) defines a local coordinate transformation Q_3 of Type II, and $F\circ(Q_3\circ Q_2\circ Q_1)^{-1}$ takes te form as in (19), Case II, $k \geq 2$.

It remains to establish Case II, $k = 1$. In this case, we put:

$$\left.\begin{array}{l} u_1 = F(x,0),\ u_j = \phi_j(x),\ j = 2,\dots,n, \\ v = \varepsilon(F(x,y) - F(x,0)). \end{array}\right\} \tag{29}$$

In (29), the C^∞-functions ϕ_j are chosen such that - besides $\phi_j(0) = 0$ - the set $\{D_xF(0), D\phi_j(0), j = 2,\dots,n\}$ is linearly independent. Obviously, (29) defines a local coordinate transformation Q of Type II and $F\circ Q^{-1}(u,v) = \varepsilon v +$

u_1, which corresponds to (19), Case II, $k = 1$. This completes the proof of Lemma 1. □

Remark 6. The result in Lemma 1, Case II, can also be derived by means of the ideas in the paper of Siersma (cf. [13]). The coordinate transformation of Type II is used to obtain local diffeomorphisms of the set $\{y \mid y \geq 0\}$ to itself. Roughly speaking, in order to avoid the inequality "$y \geq 0$", Siersma's nice idea was to replace y by y^2 and to introduce a symmetry-argumentation.

Remark 7. Concerning the proof of Lemma 1 we emphasize the following (which follows from (22), (23)).
In Case I, we have: $\text{span}\{D\alpha_i(0),\ i = 1,..,k\} = \text{span}\ S(0)$, (cf. (6)).
In Case II, $k \geq 2$, we have (recall the "replacement of k by $k-1$"):
$\text{span}\{D\alpha_i(0),\ i = 1,...,k-1,\ D_w\beta(0)\} = \text{span}\ S(0)$.

Remark 8. Note that the choice of the functions ϕ_j appearing in (25), (26), (29), admits a relatively great amount of freedom.

Proof of Theorem 1. A moment of reflection shows that for the proof it remains to make a combination of Lemma 1, Remark 7, Remark 8, noting that ε in (17), (18) is always +1 (since we are dealing with (global) minima in the y-coordinate; in case $y = b$, the orientation of y has to be reversed). □

Proof of Theorem 2.

We shall introduce two subsets of $C^\infty(R^{n+1},R)$, namely $\mathcal{F}^*$ and $\mathcal{F}^{**}$, both of which are C^k_S-dense for all k, and C^k_S-open for all $k \geq n+1$. Moreover $\mathcal{F}^{**} \subset \mathcal{F}^*$ and $\mathcal{F}^{**} \subset \mathcal{F}$, which shows the dense-part of Theorem 2. For all this, we also need to introduce two sets $\mathcal{H}$, $\mathcal{G}$, which will have a natural Whitney-stratification. The latter fact enables us to exploit a "general position"-argument.

Definition 6. An $F \in C^\infty(R^{n+1},R)$ is said to have regular zero-points w.r.t. $[a,b]$, if the following holds, whenever $F(x,y) = 0$ and $y = a,b$, or,

$F(x,y) = \frac{\partial}{\partial y}F(x,y) = 0$ and $y \in (a,b)$: y is a zero-point for $F(x,\cdot)$ of order k, $1 \le k \le n$, and S(y) is linearly independent.
By F^* we denote the set consisting of all $F \in C^\infty(R^{n+1},R)$ having regular zero-points w.r.t. [a,b].

Definition 7. The reduced (n+1)-jet extension of $F \in C^\infty(R^{n+1},R)$ is defined by means of the following map j_F:

$$j_F(x,y) = (x,y,F(x,y), \frac{\partial}{\partial y} F(x,y),\ldots,\left(\frac{\partial}{\partial y}\right)^{n+1} F(x,y)). \tag{30}$$

The image space of j_F is called the reduced (n+1)-jet space J.

Remark 9. The space J is equal to R^{2n+3}. Moreover, the set $j_F(R^{n+1})$ is an (n+1)-dimensional C^∞-manifold in J.

Lemma 2. The set F^* is C_S^k -dense for all k. Moreover, F^* is C_S^k -open for all $k \ge n+1$.

Proof. Let 0_p denote the origin in R^p. We define the following manifolds in J:

$$\left.\begin{aligned} M^i &= R^n \times (a,b) \times \{0_{i+1}\} \times R^{n+1-i}, \ i = 1,\ldots,n+1, \\ M^i_c &= R^n \times \{c\}\times\{0_i\} \times R^{n+2-i}, \ i = 1,\ldots,n+2, \text{ where } c = a,b. \end{aligned}\right\} \tag{31}$$

Obviously, M^i, M^i_a, M^i_b have codimension i+1 in J. Let T denote the set consisting of all $F \in C^\infty(R^{n+1},R)$ for which $j_F(R^{n+1})$ intersects M^i, M^j_a, M^j_b, $i = 1,\ldots,n+1$, $j = 1,\ldots,n+2$, transversally. From Thom's transversality theory (cf. [1], [10]) it follows that T is C_S^k -dense for all k. Next, a (co-) dimension argument shows : if $F \in T$, then $j_F(R^{n+1})$ does not intersect M^i, M^i_a, M^j_b, for $i = n+1$, $j = n+1, n+2$. Keeping in mind that a and b are the boundary points of the open interval (a,b), a continuity argument shows that T is C_S^k -open for all $k \ge n+1$. Finally, writing down the very transversality conditions in the definition of T, we see that $F^* = T$, which proves the lemma. □

Definition 8. An $F \in C^\infty(R^{n+1},R)$ is said to have regular zero-points (w.r.t. [a,b]) <u>in general position</u> if $F \in F^*$ and if for every $x \in R^n$ the finite (possibly empty) set of zero-points $y_1,\dots,y_p \in [a,b]$, $1 \le \mathrm{Ord}(y_i) \le n$, satisfies the following additional condition:

$$\dim \operatorname{span} \Big[\bigcup_{i=1}^{p} S(y_i) \Big] = \sum_{i=1}^{p} \mathrm{Ord}(y_i). \tag{32}$$

By F^{**} we denote the elements of F^{*} satisfying the above additional condition.

Lemma 3. The set F^{**} is C^k_S -dense for all k. Moreover, the set F^{**} is C^k_S-open for all $k \ge n+1$.

Remark 10. Note that the set F^{**} is contained in the set F defined by (12). Hence, Lemma 3 implies the dense-part of Theorem 2.

In the following definition we follow [6].

Definition 9. Let W be a subset of R^k and $\mathcal{W}$ a locally finite partition of W into C^∞-manifolds (each manifold having a specific dimension). The family $\mathcal{W}$ is called a stratification of W and each element of $\mathcal{W}$ is called a stratum.
Let $X, Y \in \mathcal{W}$ and $x \in X$. Then, Y is called Whitney regular over X at x if the following holds: if (x^i), (y^i) are sequences in X, Y, $(x_i \neq y_i)$, both converging to x, if the sequence of tangentspaces $T_{y_i}Y$ converges to a subspace $T \subset R^k$ (in the corresponding Grassmannian) and if the sequence $(\overrightarrow{x^i y^i})$ of lines containing $x_i - y_i$ converge to a line $L \subset R^k$ (in the projective space), then $L \subset T$.
We say that Y is Whitney regular over X when it is so at every point in X.
Moreover, $\mathcal{W}$ is called a Whitney stratification of W if every stratum $Y \in \mathcal{W}$ is Whitney regular over any other stratum $X \in \mathcal{W}$.

Remark 11. Note that Whitney regularity is preserved under (local) diffeomorphisms of the embedding space R^k.
A Whitney regular stratification $\mathcal{W}$ of $W \subset R^k$ has the following special property which allows "general position" arguments: if M is a C^∞-manifold intersecting a stratum $X \in \mathcal{W}$ at $x \in X$ transversally, then M intersects, locally around x, all those strata of W transversally, which contains $\bar{x}$ in their closure.

In the spirit of Lemma 1, we introduce, for $k \geq 1$, $\varepsilon \in \{-1,+1\}$, the following typical functions:

$$H(x,y) = \varepsilon y^{k+1} + \sum_{i=1}^{k} x_i y^{k-i} \quad , \quad G(x,y) = \varepsilon y^{k} + \sum_{i=1}^{k} x_i y^{k-i}. \tag{33}$$

The functions H, G in (33) generate the following subsets of R^k:

$$\mathcal{H} = \{x \in R^k \mid \exists y \text{ such that } H(x,y) = 0, \frac{\partial}{\partial y} H(x,y) = 0\}, \tag{34}$$

$$\mathcal{G} = \{x \in R^k \mid G(x,0) = 0 \text{ or, } \exists\, y > 0 \text{ such that } G(x,y) = 0, \frac{\partial}{\partial y} G(x,y) = 0\}. \tag{35}$$

Let $x \in \mathcal{H}$. Then, there exist unique integers q_i, $i = 1,\dots,r$, with $1 \leq q_1 \leq \dots \leq q_r$, and distinct real numbers α_i, $i = 1,\dots,r$, such that

$$H(x,y) = p(y) \prod_{i=1}^{r} (y-\alpha_i)^{q_i+1} \tag{36}$$

where p(y) is a polynomial in y, $p(\alpha_i) \neq 0$, all i, whose zeros are either real, simple, or complex. The set of points $x \in \mathcal{H}$ for which (36) holds is denoted by $\mathcal{M}(q_1,\dots,q_r)$.

Let $x \in \mathcal{G}$. Then, there exist unique integers q_i, $i = 0,1,\dots,r$, with $q_0 \geq 0$, $1 \leq q_1 \leq \dots \leq q_r$, and distinct real numbers α_i, $i = 1,\dots,r$, such that

$$G(x,y) = p(y).y^{q_0}. \prod_{i=1}^{r} (y-\alpha_i)^{q_i+1} , \tag{37}$$

where p(y) is a polynomial with $p(0) \neq 0$, $p(\alpha_i) \neq 0$, all i, whose zeros are either real, negative, or complex, or real, positive, simple. The set of points $x \in \mathcal{G}$ for which (37) holds is denoted by $\mathcal{N}q_0(q_1,\dots,q_r)$, resp. $\mathcal{N}q_0$ if no q_i, $i \geq 1$, appears in (37).

In Fig. 1 the sets $\mathcal{H}, \mathcal{G}$ are depicted for $k = 3$, $\varepsilon = 1$. In this case, $\mathcal{H}$ is the socalled "swallow-tail" (cf. [4]).

<u>Lemma 4.</u>

a. Each set of the type $\mathcal{M}(q_1,\dots,q_r)$, resp. $\mathcal{N}q_0(q_1,\dots,q_r)$, $\mathcal{N}q_0$, is a

C^{∞}-manifold in R^k of codimension $\sum_{i=1}^{r} q_i$, resp. $\sum_{i=0}^{r} q_i$, q_o.

b. The sets H, G are closed subsets of R^k.

c. The manifolds $M(q_1,\ldots,q_r)$ (resp. $N_{q_o}(q_1,\ldots,q_r)$, N_{q_o}) form a Whitney stratification for H (resp. G).

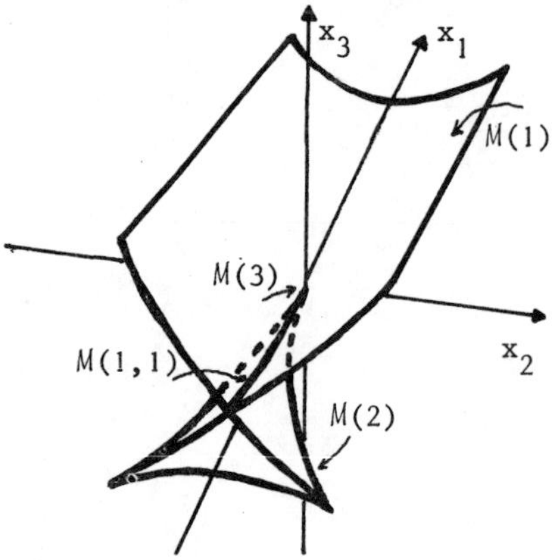

Fig.1. The set H, k=3, ε=1.

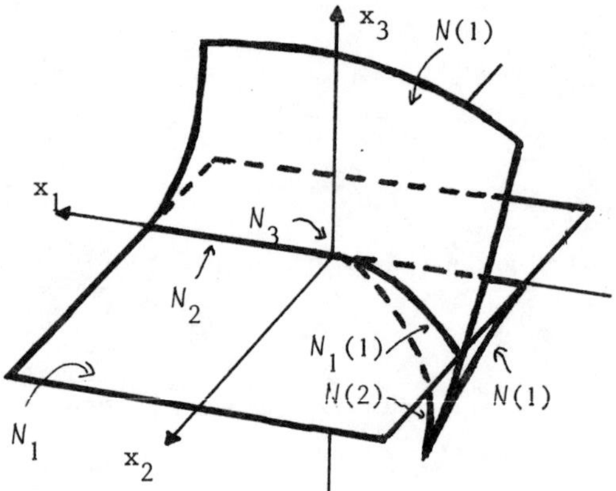

The set G, k=3, ε=1.

For the proof of Lemma 4 we need the following technical lemma, the proof of which will be omitted, since it consists of straightforward elementary calculations.

<u>Lemma 5.</u> For fixed $n \geq 0$ put $\rho(y) = (y^n, y^{n-1},\ldots,y,1)$. Given $r \geq 1$, let $\kappa_i \geq 0$, $i = 1,\ldots,r$, be integers satisfying the inequality $\sum_{i=1}^{r} (\kappa_i+1) \leq n+1$.

a. Let α_i, $i = 1,\ldots,r$, be distinct real numbers. Then, the vectors $\left(\frac{\partial}{\partial y}\right)^j \rho(\alpha_i)$, $j = 0,1,\ldots,\kappa_i$, $i = 1,\ldots,r$, are linearly independent.

b. For fixed $k \geq 1$, let $\alpha_i^{(k)}$, $i = 1,\ldots,r$, be distinct real numbers, such that $\alpha_i^{(k)} \to 0$ for all i as $k \to \infty$. Let $M_k \subset R^{n+1}$ be the linear subspace orthogonal to the vectors $\left(\frac{\partial}{\partial y}\right)^j \rho\,(\alpha_i{}^{(k)})$, $j = 0,1,\ldots,\kappa_i$, $i = 1,\ldots,r$. (Note: in view of Statement a, M_k is a linear subspace of R^{n+1} of codimension $\sum_{i=1}^{r} (\kappa_i+1)$). Then M_k converges to a linear subspace M, in the corresponding Grassmannian, as $k \to \infty$.

Proof of Lemma 4.

Statement a. Let $\bar{x} \in \mathcal{M}(q_1,\dots,q_r)$, let $\bar{\alpha}_i$, $i = 1,\dots,r$, be distinct real numbers according to (36) and consider, for each i, the set of equations:

$$\left(\frac{\partial}{\partial y}\right)^j H(x,y) = 0,\ j = 0,1,\dots,q_i\ (i=1,\dots,r) \tag{38}$$

For each fixed i, $(\bar{x},\bar{\alpha}_i)$ satisfies (38). From (33) it follows that the vectors $\left(\frac{\partial}{\partial y}\right)^j D_x H(\bar{x},\bar{\alpha}_i)$, $j = 0,1,\dots,q_i-1$, are linearly independent. Moreover, $\left(\frac{\partial}{\partial y}\right)^{q_i+1} H(\bar{x},\bar{\alpha}_i) \neq 0$. It follows that the $(q_i+1)\times(k+1)$matrix with $D\left(\frac{\partial}{\partial y}\right)^{\ell-1} H(\bar{x},\bar{\alpha}_i)$ as its ℓ-th row, has the form:

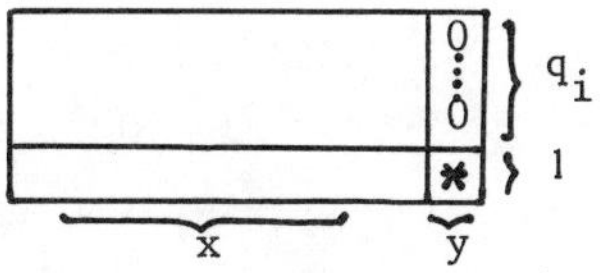

Hence, in view of the implicit function theorem, in an open neighborhood $\mathcal{Q}$ of $(\bar{x},\bar{\alpha}_i)$, we can express both α_i and "q_i of the components $(x_1,\dots,x_k)$" as C^∞ -functions of the remaining "k-q_i components of $(x_1,\dots,x_k)$" such that (38) is satisfied identically. In this way we obtain, locally, a manifold $\mathcal{M}^i_{loc}$ in x-space R^k of codimension q_i with $\left(\frac{\partial}{\partial y}\right)^j D_x H$, $j = 0,1,\dots,q_i-1$, as a set of normal vectors. Using Lemma 5.a, we see that the set of vectors

$$\left(\frac{\partial}{\partial y}\right)^j D_x H,\ j = 0,1,\dots,q_i-1,\ i = 1,\dots,r \tag{39}$$

is linearly independent. Hence, the manifolds $\mathcal{M}^i_{loc}$, $i = 1,\dots,r$, are in general position at $\bar{x}$. But, locally, $\mathcal{M}(q_1,\dots,q_r)$ is the intersection of the manifolds $\mathcal{M}^i_{loc}$. This proves the fact that $\mathcal{M}(q_1,\dots,q_r)$ is a manifold in R^k of codimension $\sum_{i=1}^{r} q_i$. For $\mathcal{N}q_0(q_1,\dots,q_r)$, resp. $\mathcal{N}q_0$, a similar argumentation can be used, where, in case i = 0, the set of equations according to (38)is the following:

$$\left(\frac{\partial}{\partial y}\right)^j H(x,0) = 0\ ,\ j = 0,1,\dots,q_0-1. \tag{40}$$

Statement b. It is not difficult to recognize that the set

$\overline{M(q_1,\ldots,q_r)}\setminus M(q_1,\ldots,q_r)$ consists of strata $M(\ldots)$ of higher codimension. From this it follows that H is closed. For the set G a similar argumentation holds.

Statement c. We start with the following remark, resulting from the argumentation in te proof of Statement b : it suffices to cheque that the higher dimensional strata are Whitney regular over the lower dimensional strata.

The set H. We use induction on the dimension k (cf. (34)), the case k =1 being trivial. To this aim we use the notation H_k instead of H. Firstly, we consider the origin $0 \in H_k$. Note that $M(k) = \{0\}$ is the unique lowest dimensional stratum of H_k. Let (x^n) be a sequence in $M(q_1,\ldots,q_r)$, $\sum_{i=1}^{r} q_i < k$, converging to the origin. If $x^n \to 0$, then the corresponding real numbers α_i^n, $i = 1,\ldots,r$, in (36) converge to zero. The set of vectors in (39), evaluated (for each i) in (x^n,α_i^n) form a set of normal vectors for $M(q_1,\ldots,q_r)$ at x^n. Then, from (33) and Lemma 5b it follows that the tangentspaces $T_{x^n} M(q_1,\ldots,q_r)$ converge to a limit, say $T(q_1,\ldots,q_r)$. From (38) and Lemma 5a it is not difficult to see that each accumulation point of the sequence $\overrightarrow{(0\ x^n)}$ of lines in $T(q_1,\ldots,q_r)$. This implies the Whitney regularity of $M(q_1,\ldots,q_r)$ over $\{0\}$. Now, let $\bar{x} \in H_k\setminus\{0\}$, say $\bar{x} \in M(q_1,\ldots,q_r)$. Using the idea of the proof of Lemma 1, Case I, it follows that there exists a neighborhood Q of $\bar{x}$ such that $H_k \cap Q$ is the union of r sets, each of them locally diffeomorphic to a neighborhood of the origin of $H_{q_i} \times R^{k-q_i}$, $i = 1,\ldots,r$. By induction, H_{q_i} has an obvious Whitney stratification, and a stratum of $H_{q_i} \times R^{k-q_i}$ has the form $M \times R^{k-q_i}$, where M is a stratum of H_{q_i}. It is easily seen (cf. Lemma 5a) that the strata of the locally diffeomorphic copies of $H_{q_i} \times R^{k-q_i}$, for different i's, are in general position. This, finally, implies the Whitney regularity of the stratification of H_k.

The set G. The proof of the Whitney regularity of the given stratification of G runs in a simular way as is done in case of the set H. The only difference is that the idea of the proof of Lemma 1 w.r.t. to both Case I and Case II has to be exploited. □

Proof of Lemma 3. Firstly, we note that the open-part of Lemma 2 allows appropriate perturbations of an $F \in \mathcal{F}^*$, thereby staying within $\mathcal{F}^*$. The key-idea is to chose these perturbations in a sophisticated way. In fact, pick an $\bar{x} \in \mathbb{R}^n$ and suppose that the set of zero-points of order k, $1 \leq k \leq n$, is not empty, say $\{\bar{y}_1,\ldots,\bar{y}_p\}$. Put $k_i = \mathrm{Ord}(\bar{y}_i)$, $i = 1,\ldots,p$. Following the idea of Lemma 1 we see that there exists a neighborhood $\mathcal{Q}$ of $\bar{x}$ such that the subset of $\mathcal{Q}$ consisting of those x for which the corresponding zero-set is not empty, is the union of locally diffeomorphic images of the sets $\mathcal{H}_{k_i} \times \mathbb{R}^{n-k_i}$, resp. $\mathcal{G}_{k_i} \times \mathbb{R}^{n-k_i}$ (cf. (34), (35), and replace thereby the dimension k by k_i). From Lemma 4 we know that each of these sets has a natural Whitney stratification. Now the fact is that these stratified sets need not to be in general position to each other. But this is easily accomplished by means of suitable local perturbations of F by means of polynomials, thus "moving" the stratified sets independently from each other, using Sard's theorem and taking Remark 11 into account. Once our stratified sets are in general position (locally), this situation is locally stable in view of Lemma 4.b. Altogether, this roughly proves Lemma 3. □

In view of Remark 10, it remains to show the open-part of Theorem 2. In fact, this will be settled by the following. Suppose $\phi_F(\bar{x}) = 0$ and $E_F(\bar{x}) = \{\bar{y}_1,\ldots,\bar{y}_p\}$ according to Definition 4, R1. Obviously, the maximal order of an $\bar{y}_i$ equals n. Firstly, assume that $\mathrm{Ord}(\bar{y}_1) = n$. Then, $E_F(\bar{x})$ is a singleton, say $\{\bar{y}\}$. Since $\bar{y}$ is a minimum for $F(\bar{x},\cdot)|_{[a,b]}$ this situation cannot occur if $\bar{y} \in (a,b)$ and n is even. This clarifies the distinction between the cases n is odd, resp. n even, in the open-part of Theorem 2. Next,it is easily seen (again following Lemma 1 and using a continuity argument) that there exists an open neighborhood $\mathcal{Q}$ of $\bar{x}$ such that the set of those $x \in \mathcal{Q}$ for which there exists an $y \in [a,b]$ being a zero-point of order k, $1 \leq k \leq n$, consists of the union of diffeomorphic images of sets $\mathcal{H}_{k_i} \times \mathbb{R}^{n-k_i}$, $i = 1,\ldots,p$, where $k_i = \mathrm{Ord}(\bar{y}_i)$, being in general position. Using again Lemma 4.b, as exploited in the proof of Lemma 3, and using the very definition of the C^k_S -topology, this completes the proof of the open-part of Theorem 2. □

In the case n = 3, $\phi_F(\bar{x}) = 0$, $E_F(\bar{x}) = \{\bar{y}\}$, $\mathrm{Ord}(\bar{y}) = 3$, the set M as intruduced in (8) has locally -up to diffeomorphism- the form of the "upper halfspaces" bounded by the sets depicted in Fig. 2 (cf. also Fig. 1).

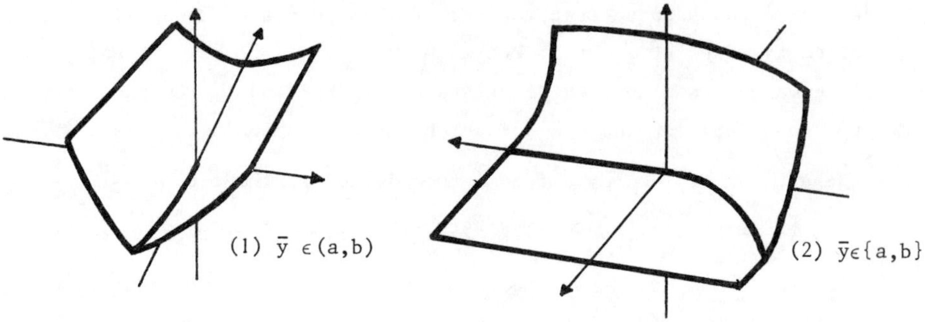

Fig.2. The set $\psi_F = 0$ in case: $E_F(\bar{x}) = \{\bar{y}\}$, n=3, Ord$(\bar{y})$=3.

In Fig. 3 we have depicted the change of F(x,•) as x describes a loop around the origin w.r.t. the boundary-sets in Fig. 2. (1), (2).

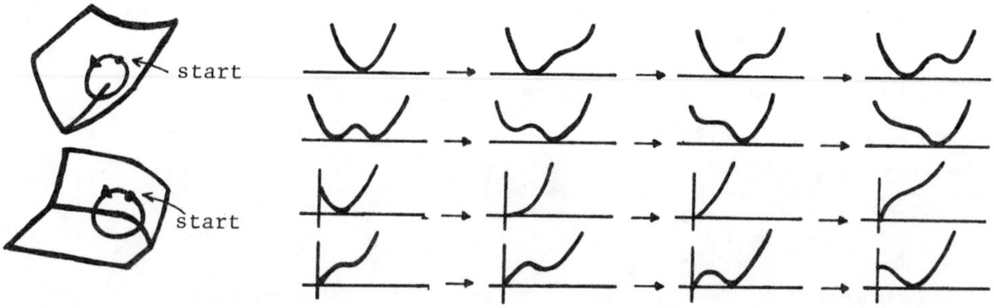

Fig.3. The change of F(x,.) as x describes a loop around the origin.

Final remarks.

We mention a number of important contributions, which are related to our investigations. In [2] Brizgalova studies (normal forms of) critical points of functions of maximum type, where the maximum is taken over a compact smooth manifold without boundary and where the number of parameters does not exceed six. In [3] an inequality is presented which estimates the number of essential dimensions being necessary for the local description of a function of maximum type in relation with the number of parameters. In [12] Matov considers the (local) topological equivalence of a function of maximum type (and even minimax type) with a Morse-function.

References.

[1] Арнольд,В.И., Варченко,А.,Н., Гусейн-Заде,С.М.(1982) Особенности дифференцируемых отображений.Издательство Наука, Москва.

[2] Брызгалова,Л.Н.(1977) Особенности максимума функции, зависящей от параметров. Функц.Анализ, т.11, вып.1, с.59-60.

[3] Брызгалова,Л.Н.(1978) О функциях максимума семейства функций, зависящих от параметров. Функц.Анализ, т.12, вып.1, с.66-67.

[4] Bröcker,Th., Lander,L.(1975) Differentiable Germs and Catastrophes. London Math.Soc. Lect.Note Series 17, Cambridge University Press.

[5] Fujiwara,O.(1983) A Note on Differentiability of Global Optimal Values. Preprint, Asian Institute of Technology, Bangkok, Thailand.

[6] Gibson,C.G., Wirthmüller,K., du Plessis,A.A., Looijenga,E.J.N.(1976) Topological Stability of Smooth Mappings. Lect.Notes Math., Vol.552, Springer Verlag.

[7] Hettich,R., Jongen,H.Th.(1978) Semi-infinite Programming: Conditions of Optimality and Applications. In: Optimization Techniques, Part 2, Lect. Notes Control and Inf.Sc., Vol.7,Springer Verlag, pp.1-11.

[8] Hettich,R.,Jongen,H.Th.(1981) On the Local Continuity of the Chebyshev Operator. J.Approx.Theory, Vol.33, No.4, pp.296-307.

[9] Hettich,R.,Zencke,P.(1982) Numerische Methoden der Approximation und semi-infiniten Optimierung. Teubner Studienbücher, Stuttgart.

[10] Hirsch,M.W.(1976) Differential Topology. Springer Verlag.

[11] Jongen,H.Th., Jonker,P., Twilt,F.(1983) Nonlinear Optimization in $\mathbb{R}^n$, I.Morse Theory,Chebyshev Approximation. Methoden und Verfahren der Mathematischen Physik, Band 29, Peter Lang Verlag, Frankfurt a.M.

[12] Матов,В.И.(1982) Топологическая классификация ростков функций максимума и минимакса семейств функций общего положения. УМН, т.37, вып.4, с.167-168.

[13] Siersma,D.(1981) Singularities of Functions on Boundaries, Corners, etc. Quart.J.Math., Oxford Ser.(2) 32, no.125, pp.119-127.

[14] Wetterling,W.(1970) Definitheitsbedingungen für relative Extrema bei Optimierungs- und Approximationsaufgaben. Numer.Math.15, pp.122-136.

Author's address

Prof.Dr.H.Th.Jongen, Ir.G.Zwier,
Twente University of Technology
Dept.Applied Mathematics
P.O.Box 217
7500AE Enschede
The Netherlands

International Series of
Numerical Mathematics, Vol. 72

MOST OF THE OPTIMIZATION PROBLEMS HAVE UNIQUE SOLUTION

Petar S. Kenderov

Institute of Mathematics
Sofia, Bulgaria

1. Summary

Let X be a compact metric space and C(X) be the space of all continuous real-valued functions in X. Every pair (A,f), where A belongs to the set 2^X of all closed subsets of X and f is from C(X), determines a (constrained) minimization problem: $\min\{f(y) : y \in A\}$ (find x A at which f attains its minimum over A). Suppose that 2^X is endowed with the Hausdorff metric and C(X) is topologized by the usual uniform convergence norm. We prove that there is a dense G_δ-subset G of 2^X C(X) such that every minimization problem (A,f) from G has unique solution, i.e. the set $\{x \in A : f(x) = \min\{f(y) : y \in A\}\}$ consists of only one point for each pair (A,f) outside some first Baire category subset of $2^X \times C(X)$.

There are nonmetrizable compacts X for which this result is still valid. If X is a weakly compact subset of a Banach space or if X is homeomorph to a weak*-compact subset of a dual Banach space with the Radon-Nikodym property, then the result is also true. Moreover, the method of proof shows that this result has place for all compacts from a large class of spaces, which was introduced recently by Stegall. This class is closed under taking countable products, countable sums of closed sets and perfect images.

2. Definitions, Notations, and Main Results

Let X be an arbitrary compact space. The exponential topology (called also Vietoris topology) in 2^X has a base consisting of sets of the type $\{ A \in 2^X : A \subset U_0 , A \cap U_i \neq \emptyset, i = 1, 2, \ldots, k \}$, where $U_0, U_1, \ldots, U_k$, are arbitrary open subsets of X. It is well known that 2^X with this topology is also compact. In the particular case when X is a metric compact space, the exponential topology in 2^X coincides with the topology generated by the Hausdorff metric in 2^X (see, for instance, the book of Kuratowski [11]).

The multivalued mapping $F: Z \rightarrow X$ assigning to each z from the topological space Z some subset F(z) of X is said to be upper semi-continuous (lower semicontinuous) at some point $z_0 \in Z$ if for every open $U \subset X$, $F(z_0) \subset U$ $(F(z_0) \cap U \neq \emptyset)$, there exists an open $V \subset Z$, $V \ni z_0$, such that $F(z) \subset U$ $(F(z) \cap U \neq \emptyset)$ whenever $z \in V$. When F is upper (lower) semi-continuous at each point $z \in Z$ we simply say that F is usc (lsc) in Z. Mappings which are simultaneously usc and have compact nonempty images F(z) for every $z \in Z$ are called "usco mappings" (see Christensen [1]).

Let us now consider the multivalued mapping S acting from $2^X \times C(X)$ into X according to the rule: for $A \in 2^X$ and $f \in C(X)$ $S(A,f) = \{ x \in A : f(x) = \min \{ f(y) : y \in A \} \}$, i.e. to each pair $(A,f) \in 2^X \times C(X)$ the mapping S puts into correspondence the (nonempty) set of all solutions to the minimization problem "$\min \{ f(y) : y \in A \}$" determined by (A,f).

<u>Proposition 1</u>. The just defined mapping $S: 2^X \times C(X) \rightarrow X$ is usco.

Proof. Since X is compact it is enough to show that the graph $G(S) = \{ (A,f,x) \in 2^X \times C(X) \times X : x \in S(A,f) \}$ is a closed set in $2^X \times C(X) \times X$.

Let the net $\{(A_\alpha, f_\alpha, x_\alpha)\}_\alpha$ converge to (A_0, f_0, x_0) and $x_\alpha \in S(A_\alpha, f_\alpha)$ for each α. We will show that $x_0 \in S(A_0, f_0)$. The latter is equivalent to the following two statements

a) $x_0 \in A_0$

b) $f_0(x_0) \leqq f_0(y)$ for every $y \in A_0$.

a) By the convergence of A_α to A_0 in 2^X each open in X neighbourhood U of A_0 will contain A_α when α is large enough. As $x_\alpha \in A_\alpha$ the closure $\overline{U}$ of U in X will contain x_0. Since the space X is regular (even normal), this is enough to conclude that $x_0 \in A_0$.

The proof of b) is also a routine matter. Let $y_0 \in A_0$. Since $\{A_\alpha\}_\alpha$ converges to A_0, there exists a net $\{y_\alpha\}_\alpha$, $y_\alpha \in A_\alpha$, having y_0 as a cluster point. Taking a subnet (if necessary) we may presume that $\{y_\alpha\}_\alpha$ converges to y_0. Take $\varepsilon > 0$. For sufficiently large α , we have

(1) $$|f_0(x_0) - f_0(x\ )| < \varepsilon$$

(2) $$\|f_\alpha - f_0\| < \varepsilon \text{ , and}$$

(3) $$|f_0(y_0) - f_0(y_\alpha)| < \varepsilon \ .$$

Besides, for each α

(4) $$f_\alpha(x_\alpha) \leqq f_\alpha(y_\alpha) \quad (\text{because } x_\alpha \in S(A_\alpha, f_\alpha)) \ .$$

Due to (1), (2), (3) and (4), for large α , the following chain of inequalities holds:

$$f_0(x_0) \leqq f_0(x\ ) + \varepsilon \leqq f_\alpha(x_\alpha) + 2\varepsilon \leqq$$

$$f_\alpha(y_\alpha) + 2\varepsilon \leqq f_0(y_\alpha) + 3\varepsilon \leqq f_0(y_0) + 4\varepsilon \ .$$

Since $\varepsilon > 0$ is arbitrarily chosen, the claim b) is proved. This completes the proof of Proposition 1.

For our considerations we will need one more type of continuity of multivalued mappings, which will be called "property (∗)". In [5] this notion is related to the existence of approximate continuous selections to multivalued maps.

Definition 1 (property (∗), see [1,2]). The mapping $F:Z \longrightarrow X$ is said to have the property (∗) at the point $z_o \in Z$ if there exists such a point $x_o \in F(z_o)$ that for every open $U \subset X$, $U \ni x_o$, an open $V \subset Z$, $V \ni z_o$, exists for which $F(z) \cap U \neq \emptyset$ whenever $z \in V$.

Evidently, if F is lsc at some point z_o, then it has the property (∗) at z_o. Also, any multivalued mapping $F:Z \longrightarrow X$ admitting a single-valued selection which is continuous at some z_o, also has the property (∗) at z_o.

Let $\mathfrak{K}$ be a class of topological spaces. Starting from $\mathfrak{K}$ another class $\mathcal{L}(\mathfrak{K})$ can be determined.

Definition 2. We will say that a given topological space X belongs to the class $\mathcal{L}(\mathfrak{K})$ if for every space $Z \in \mathfrak{K}$ and for every usco mapping $F:Z \longrightarrow X$ the set of points where F does not have the property (∗) is of the first Baire category in Z.

The next result shows that all metric spaces belong to $\mathcal{L}(\mathfrak{K})$ for every class $\mathfrak{K}$ (the theorem is, if fact, a much stronger statement). Other results of the same type can be found in [7, 8, 9, 10].

Theorem 1 (Fort [6]). Let X be a metrizable space and $F:Z \longrightarrow X$ be an usco mapping. Then the set $\{ z \in Z : F \text{ is not lsc at } z\}$ is of the first Baire category in Z.

Further general properties of the class $\mathcal{L}(\mathfrak{K})$ will be discussed in the next section of this paper. Now we will consider the particular case when the class $\mathfrak{K}$ consists of all Čech-complete spaces. In this case we will write $\mathcal{L}$ instead of

$\mathcal{L}(\mathfrak{A})$.Let us remind that the topological space Z is said to be Čech-complete if it is homeomorph to a G_δ-subset of some compact space.It is known that complete metric spaces and locally compact spaces are Čech-complete.Every space $Z = Z_1 \times Z_2$, where Z_1 is a compact space and Z_2 is a complete metric space (or, more generally, Čech-complete space) is again Cech-complete space.In particular, for arbitrary compact space X the space $2^X \times C(X)$ is Čech- complete.All Čech-complete spaces are of the second Baire category and the intersection $\bigcap_{i=1}^{\infty} U_i$ of countably many open and dense subsets U_i of a given Čech-complete space Z is still dense in Z.

Each Banach space E with its weak topology is in $\mathcal{L}$ (Christensen [1]).If a dual Banach space E^* has the Radon-Nikodym property, then E^* equipped with the weak* topology is in $\mathcal{L}$ (Christensen and Kenderov [2]).In fact, in the last two mentioned papers something more is proved.For every usco mapping $F:Z \longrightarrow (E, \text{weak})$ ($F:Z \longrightarrow (E, \text{weak}^*)$), where Z is Čech-complete and E is a Banach space (E has the Radon-Nikodym property), there exists a dense G_δ-subset of Z at the points of which F has the property (✱) with respect to the norm topology in E (E^*).

<u>Theorem 2</u>. Let X be a compact space from the class $\mathcal{L}$ and 2^X be endowed with the exponential topology,Consider the above defined mapping $S:2^X \times C(X) \longrightarrow X$ assigning to each pair (A,f) from $2^X \times C(X)$ the set $S(A,f) = \{x \in A : f(x) = \{\min f(y) : y \in A\}\}$. Then there exists a dense G_δ-subset G of $2^X \times C(X)$ such that S(A,f) is a singleton for each (A,f) from G (i.e. "almost all" minimization problems (A,f) have unique solution).

Proof. By Proposition 1 the mapping S is usco.Since $X \in \mathcal{L}$ and $2^X \times C(X)$ is Čech-complete, there exists a dense G_δ-subset G of $2^X \times C(X)$ at each point of which S has the property (✱).It remains to prove that for each point (A,f), where the requirements of the property (✱) are fulfilled , the set S(A,f) has only one point.Let (A_0,f_0) be a point at which S has the property (✱) and let $x_0 \in S(A_0,f_0)$ be the point from the definition of property (✱).Suppose that $S(A_0,f_0)$ contains some other point

$x_1 \neq x_0$. By the compactness of X we find two disjoint open neighbourhoods U_0, U_1, $x_i \in U_i$, $i = 0,1$ and a nonnegative function $g \in C(X)$ such that $g(x) = 1$ for $x \in U_0$ and $g(x) = 0$ for $x \in U_1$. Consider the function $f_t(x) = f_0(x) + tg(x)$, $t > 0$. As $t > 0$ tends to 0, the net $(A_0, f_t)_t$ converges to (A_0, f_0) in the topology of $2^X \times C(X)$. Nevertheless $S(A_0, f_t) \cap U_0 = \emptyset$. This contradicts the property $(*)$ and the proof of Theorem 2 is completed.

Theorem 3. Let X be a compact space from $\mathcal{L}$ and X_1 be a closed subset of X, There exists a dense G_δ-subset G of C(X) such that each function from G attains its minimal (maximal) value in X_1 at only one point.

Proof. Consider the mapping $S_1 : C(X) \longrightarrow X_1$ acting in accordance with the rule $S_1(f) = \{ x \in X_1 : f(x) = \min\{f(y) : y \in X_1\}\}$, i.e. $S_1(f) = S(X_1, f)$. The mapping S_1 is usco. As above it is poosible to prove that at the "points" $f \in C(X)$ where S_1 has the property $(*)$ the set $S_1(f)$ consists of only one point.

Remark. The mapping $S_1 : C(X) \longrightarrow X_1 \subset (C(X))^*$ is of monotone type. As shown in [2,3] such mappings must be single-valued at the points where the property $(*)$ is fulfilled.

Corollary 1. Let X be a compact space from the class $\mathcal{L}$. Then the set of points which are G_δ-subsets of X is dense in X.

Proof. Let U be a nonempty open subset of X. Since the set of points where a function $g \in C(X)$ attains its minimum in X is a G_δ-set it is sufficient to show that there exists a function $g_0 \in C(X)$ such that $S(X, g_0)$ consists of only one point and lies in U. To do this we take a point $x_0 \in U$ and a nonnegative function $f_0 \in C(X)$ for which $f_0(x_0) = 0$ and $f_0(x) = 1$ whenever $x \in X \setminus U$. Evidently $S(X, f_0) \subset U$. Since S is usco mapping, there exists an open subset V of C(X) such that $S(X, f) \subset U$ for each f from V. Applying Theorem 3 (for the case $X_1 = X$) we find a function $g_0 \in V$ which attains its minimum at only one point $S(X, g_0) \in U$. The proof is complete.

Corollary 2. Each compact space X from the class $\mathcal{L}$ is is sequentially compact (i.e. for each sequence $\{x_i\}_{i=1}^{\infty} \subset X$ there exists a convergent subsequence $\{x_{i_k}\}_{k=1}^{\infty}$).

Proof.Let $\{x_i\}_{i=1}^{\infty}$ be a sequence in X.Without loss of generality we may assume that the set $\{x_i : i = 1,2,...\}$ has infinitely many points.Consider the set $A = \{ x \in X$: every open set $U \ni x$ contains infinitely many different points of the set $\{x_i\}\}$.Evidently, A is a nonempty closed subset of X and belongs to $\mathcal{L}$ .By the previous result A contains a point x^* which is a G_δ-subset of A.Consider the set $A_1 = A \cup \{x_i : i = 1,2,...\}$ .It is closed in X.Since the set $A_1 \setminus A$ is countable, A is a G_δ-subset of A_1.Therefore x^* will be a $G_\delta$point of A_1.It is known that in such a case the point x^* has a countable base in A_1.Since x^* is a cluster point of the set $x_i : i = 1,2,...$ it is a routine matter to conclude from here that there exists a sequence $\{x_{i_k}\}_{k=1}^{\infty}$ converging to x .The proof is completed.

From this corollary it follows that every compact non-sequentially compact space X provides an example of space for which Theorem 3 is not valid.Examples of such spaces are: the Čech-Stone compactification βN of the set N of all positive integers, and any product of uncountably many coppies of the segment $[0,1]$.

Remarks. For the particular case when X is a compact matric space and $X_1 = X$ the result contained in Theorem 3 follows from the results of De Blasi and Myjak ([4] ,Theorem 4.2; see also the remark thereafter where the reader is refered to the paper of Lucchetti and Patrone [12]).As follows from [8,10] spaces having σ-countably conservative nets belong to $\mathcal{L}$ .The same is true for spaces with σ-locally countable nets (Hansel, Jane and Kenderov [7]).

3. General Properties of the Class $\mathcal{L}(\mathfrak{A})$

A class of this type was already introduced by Stegall [13] who considered the case when $\mathfrak{A}$ consists of all Baire spaces. At first glance the definition of the Stegall's class $\mathcal{S}$ given in [13] differs from the one used here but a more close look shows that both definitions have similar (if not identical) nature. In general the "size" of the class $\mathcal{L}(\mathfrak{A})$ depends on the largeness of $\mathfrak{A}$. The narrower the class $\mathfrak{A}$ is, the broader the class $\mathcal{L}(\mathfrak{A})$ is. Therefore the class $\mathcal{L}$ is eventually larger than $\mathcal{S}$. In [13] Stegall proved that his class $\mathcal{S}$ has remarkable stability properties. $\mathcal{S}$ is closed under taking subspaces, countable products, countable sums of closed sets and perfect images. The same is true for the class $\mathcal{L}$.

Proposition 2. Let $\mathfrak{A}$ be a class of topological spaces.

a) If $X \in \mathcal{L}(\mathfrak{A})$, then $X_1 \subset \mathcal{L}(\mathfrak{A})$ for every $X_1 \subset X$;

b) Let $X \in \mathcal{L}(\mathfrak{A})$ and $g: X \longrightarrow Y$ be a continuous mapping which maps closed subsets of X into closed subsets of Y and let $g^{-1}y$ be compact and nonempty for every $y \in Y$ (mappings like g are called "perfect"). Then $Y \in \mathcal{L}(\mathfrak{A})$;

c) Suppose that together with each $Z \in \mathfrak{A}$ the class $\mathfrak{A}$ contains also every open subset of Z. Then from $X = \bigcup_{i=1}^{\infty} X_i$, where all X_i are closed in X and $X_i \in \mathcal{L}(\mathfrak{A})$, $i = 1,2,...$, if follows that $X \in \mathcal{L}(\mathfrak{A})$;

d) If the spaces X_i, $i = 1,2,...$ are in $\mathcal{L}(\mathfrak{A})$, then their Cartesian product $\prod_{i=1}^{\infty} X_i$ also belongs to $\mathcal{L}(\mathfrak{A})$;

e) Let $X \in \mathcal{L}(\mathfrak{A})$ and $h: Y \longrightarrow X$ be a continuous one-to-one mapping from Y into X. Then Y is in $\mathcal{L}(\mathfrak{A})$.

Proof. There is nothing to be proved about a). To see that b) is fulfilled one needs the following remark: For every usco mapping $F: Z \longrightarrow Y$ and any perfect mapping $g: X \longrightarrow Y$ the mapping $g^{-1}.F: Z \longrightarrow X$ is also usco. If $g^{-1}.F$ has the property $(*)$

at some $z_o \in Z$, then F also has this property at z_o. c) is also a routine matter. Let $F:Z \to X$ be an usco mapping from $Z \in \mathfrak{K}$ to $X = \bigcup_{i=1}^{\infty} X_i$, where $X_i \in \mathcal{L}(\mathfrak{K})$ and X_i is closed in X for $i = 1,2,\ldots$. The set $F^{-1}X_i \setminus \operatorname{int}F^{-1}X_i$, where intB denotes the interior of B and $F^{-1}C = \{ z \in Z : Fz \cap C \neq \emptyset \}$, is a nowhere dense subset of Z. Therefore the set $Z_o = \bigcup_{i=1}^{\infty}(F^{-1}X_i \setminus \operatorname{int}F^{-1}X_i)$ is of the first Baire category. Put $W = \bigcup_{i=1}^{\infty} \operatorname{int}F^{-1}X_i$. Evidently $Z_o \cup W = Z$. c) will be proved if we show that the set $Z_1 := \{ z \in W :$ F does not have the property (*) at $z \}$ is locally (and therefore globally) of the first Baire category in W. For each $z_o \in W$ we find an open set $V \ni z_o$ and a positive integer i_o such that $Fz \cap X_{i_o} \neq \emptyset$ for every $z \in V$. The mapping $F_{i_o}:V \to X_{i_o}$, defined by the formula $F_{i_o}(z) := F(z) \cap X_{i_o}$, is usco. Since V is from $\mathfrak{K}$ the set $\{ z \in V : F_{i_o}$ does not have the property (*)$\}$ is of the first Baire category in V. Beside this it contains the set $Z_1 \cap V$. All this shows that the set Z_1 is locally (and thus globally) of the first Baire category in W. c) is proved.

d) is not completely trivial. We need the notion "minimal usco mapping" (see Christensen [1]) which is close in spirit to the so-called irredicible perfect maps.

Definition 3. The usco mapping $F:Z \to X$ is said to be minimal if its graph does not properly contain the graph of any other usco mapping from Z to X.

The graph of any usco mapping $F:Z \to X$ is closed in $Z \times X$. Also, each mapping (with nonempty images) whose graph is a closed subset of the graph of some usco mapping is usco again. By means of Zorn's lemma it follows from this that, for each usco mapping $F:Z \to X$, there exists a minimal usco mapping $F^*:Z \to X$ with $F^*z \subset Fz$ for every $z \in Z$, If F^* has the property (*) at some point $z_o \in Z$, then F also has this property at z_o. For this reason we consider only minimal usco mappings in the definition of the class $\mathcal{L}(\mathfrak{K})$.

Minimal usco mappings have interesting properties.

<u>Lemma 1</u>. If $F:Z\to X$ is a minimal usco mapping and V is an open subset of Z, then the restriction of F in V is minimal usco mapping from V to X.

<u>Lemma 2</u>. If $F:Z\to X$ is minimal usco and $F(z_0)\cap U\neq\emptyset$ for some open $U\subset X$,then in each neighbourhood of z_0 there exsists points z' for which $F(z')\subset U$ (and, therefore, $\operatorname{int}\{z\in Z : F(z)\subset U\} \neq\emptyset$).

<u>Lemma 3</u> (Christensen [1], Theorem 1). If $F:Z\to X$ is a minimal usco mapping which has the property (*) at some z_0 then the set $F(z_0)$ consists of only one point.

Proof. Let $x_0\in F(z_0)$ be the point for which the property (*) holds.Suppose there exists a point $x_1\in F(z_0)$, $x_1\neq x_0$. Take two open sets $U_0\ni x_0$ and $U\ni x_1$ so that $U_0\cap U_1 = \emptyset$ .By Lemma 2 in each neighbourhood of z_0 there exists a point z for which $Fz'\subset U_1$.Therefore $Fz'\cap U_0 = \emptyset$ and this contradicts the condition (*).

<u>Lemma 4</u>.Let $F:Z\to X$ be a minimal usco mapping and $g:X\to Y$ be a continuous single-valued map.Then $g\circ F$ is again a minimal usco mapping.

Proof. Evidently, $g\circ F$ is an usco mapping.Take some mapping $H:Z\to Y$ which is usco and $H(z)\subset g\circ F(z)$ for every $z\in Z$. We will prove that $H = g\circ F$.Put $F_1(z) := (g^{-1}\circ H(z))\cap F(z)$ for $z\in Z$. We will see first that F_1 is an usco map.If suffices to check that F_1 has a closed graph.Take nets $z_\alpha\to z_0$, $x_\alpha\to x_0$, x_α $F_1(z_\alpha)$.As F has a closed graph and $F_1(z)\subset F(z)$, we have $x_0\in F(z_0)$.Having in mind that H was usco mapping we see from $g(x_\alpha)\in H(z_\alpha)$ and $g(x_\alpha)\to g(x_0)$ that $g(x_0)\in H(z_0)$.Therefore $x_0\in g^{-1}\circ H(z_0)$.Thus $x_0\in F_1(z_0)$.By the minimality of F we get $F\equiv F_1\equiv g^{-1}\circ H$. Hence $g\circ F = H$.Lemma 4 is proved.

Let us now turn back to the proof of d).Suppose $X = \prod_{i=1}^{\infty} X_i$ is a countable product of spaces $X_i\in\mathcal{L}(\mathfrak{F})$ $i = 1,\ldots.$ Let $F:Z\to X$ be a minimal usco mapping from $Z\in\mathfrak{F}$ into X.The mappings $F_i = p_i\circ F$, where $p_i:X\to X_i$, $i = 1,2,\ldots$,are the standard projections, are minimal usco mappings (Lemma 4). Since $X_i \in \mathcal{L}(\mathfrak{H})$ the sets $D_i = \{ z \in Z$: F_i does not have the property $(*)\}$, $i = 1,2,\ldots$ are of the first Baire category in Z. Consequently $D = \bigcup_{i=1}^{\infty} D_i$ is also a first Baire category subset of Z. By Lemma 3 at each point $z_0 \in Z \setminus D$ the mappings $F_i = p_i \circ F$, $i = 1,2,\ldots$, are single-valued. Hence the set $F(z_0)$ consists of only one point. Since F is usco, the requirements of the property $(*)$ are automatically fulfilled at each $z_0 \in Z \setminus D$. This completes the proof of d).

e) One way to prove this assertion is as follows. Consider a minimal usco mapping $F: Z \longrightarrow Y$. By Lemma 4, $h \circ F$ is also minimal usco mapping from Z to X. From Lemma 3 it follows that at the points outside some first Baire category subset of Z the mapping $h \circ F$ is single-valued. At the same points F is also single-valued. This completes the proof of e).

As stated in [13], from [14,15] it follows that separable Banach spaces in the weak topology, as well as all weakly compact subsets of any Banach space, are in $\mathcal{S}$. The same is true for the class $\mathcal{L}(\mathfrak{H})$. It follows from the renorming theorem of Trojanski [16] and the next statement.

<u>Proposition 3</u>. Let $(X, \|\cdot\|)$ be a normed space with a locally uniformly rotund norm $\|\cdot\|$. Then (X,weak) is in $\mathcal{L}(\mathfrak{H})$

Proof. Suppose $F: Z \longrightarrow (X,\text{weak})$ is an usco mapping from the Baire space Z into (X,weak). We may assume that F is minimal usco.

Denote by B the closed unit ball of $(X, \|\cdot\|)$ and define $f(z) := \min\{ \|x\| : x \in Fz$. This function is semicontinuous in Z. Therefore it is continuous at the points of some G_δ dense subset Z' of Z.

A simple separation argument reveals that at every point z of continuity of f the image Fz is contained in f(z)B (if there exists $x_0 \in Fz_0$, $\|x_0\| > f(z_0)$, then x_0 and $f(z_0)B$ can be strictly separated by a hyperplane H. By the minimality of F, any neighbourhood V of z_0 must contain points z for which Fz will lie in that halfspace defined by H which does not contain $f(z_0)B$. For such z the value of $f(z)$ is greater or equal to $\|x_0\|$; which is a contradiction).

Define now the set W_i by the formula $W_i = \{z \in Z$: there exists a neighbourhood V of z such that the diameter of the set $F(V) = \bigcup \{F(z): z \in V\}$ is less than $i^{-1}\}$. Evidently, W_i is an open subset of Z. We shall prove that it is dense in Z.

Choose some z_0 from Z' and some x_0 from $F(z_0)$. It is clear that $\|x_0\| = f(z_0)$. Take a linear functional h, $\|h\| = 1$, which strongly exposes x_0 from the ball $f(z_0)B$ and consider the set $\{x \in X : h(x) > h(x_0) - \varepsilon\} \cap (f(z_0) + \varepsilon)B$. When $\varepsilon \to 0$ the diameter of this set tends to 0 since $h(x_0) = \|x_0\|$ and $f(z_0) = \|x_0\|$. By the continuity of f at z_0 there exists an open neighbourhood V of z_0 such that $F(z) \subset (f(z_0) + \varepsilon)B$ whenever $z \in V$.

On the other hand, by the minimality of F there exists an open subset V_1 of V for which $F(V_1) \subset \{x \in X: h(x) > h(x_0) - \varepsilon\}$. This means that diam $F(V_1)$ may become very small when $\varepsilon \to 0$. This completes the proof.

Institute of Mathematics, Bulgarian Academy of Sciences,
1090 Sofia, P.O.Box 373, Bulgaria

REFERENCES

1.Jens Peter Reus Christensen, Theorems of Namioka and Johnson type for upper semi-continuous and compact-valued set-valued mappings.Proc.Amer.Math.Soc.,86(1982),649-655.

2.Jens Peter Reus Christensen and Petar Kenderov.Dense strong continuity of mapping and the Radon-Nikodym property,accepted for publ.in Math.Scandinavica.

3.J.P.R.Christensen,P.S.Kenderov,Dense Frechet differentiability of Mackey continuous convex functions,Comptes Rendus Acad.Sci. bulgare ,T.36,No.6,1983,p.737-738.

4.F.S.De Blasi and J.Myjak, Some generic properties in convex and nonconvex optimization theory,preprint.Comm.Math.(to appear)

5.Frank Deutsch and Petar S.Kenderov, Continuous selections for set-valued mappings and applications to metric projections, SIAM J.Math.Anal. 14(1983),No.1,185-194.

6.M.K.Fort,Points of continuity of semi-continuous functions, Publ.Math.Debrecen, 2(1951), 100-102 .

7.R.W.Hansel,J.E.Jayne et P.S.Kenderov,Semi-continuite inférieure générique d'une multiapplication,C.R.Acad.Sci.Paris 296 (1983).

8.P.S.Kenderov,Semi-continuity of set-valued mappings with respect to two topologies,C.R.Acad.Sci.bulgare 29 (1976),15-15.

9.P.S.Kenderov,Semi-continuity of set-valued mappings,Fund.Math. 88(1975) 61-70.

10.P.S.Kenderov, Continuity-like properties of multivalued mappings,"Serdica" Bulg.Math.Publ.,Vol.9,1983,p.149-160.

11.K.Kuratowski,Topology,v.1 (1966),v.2 (1968),Academic Press , New York and London.

12.R.Lucchetti and F.Patrone,Sulla densitáe genericitá di alquni problemi di minimo ben posti,Publicazioni dell'Inst.di Matematica, Universita di Genova n.217 (1977).

13.Charles Stegall, A class of topological spaces and differentiation of functions on Banach spaces,Preprint.

14.Charles Stegall, The Radon-Nikodym property in conjugate Banach spaces,Trans.Amer.Math.Sc. 206 (1975) 213-223.

15.Charles Stegall,The Radon-Nikodym property in conjugate Banach spaces II,Trans.Amer.Math.Soc. 264 (1981)507-519.

16.S.L.Trojanski, On locally convex and differentiable norms in certain non-separable Banach spaces,Studia Math. 37(1971), 173-180.

International Series of
Numerical Mathematics, Vol. 72

A GENERALIZATION OF THE NOTION OF CONVEXITY ON THE BASIS OF CERTAIN OPTIMIZATION PROBLEMS

František Nožička

Faculty of Mathematics and Physics,
Charles University, Praha, ČSSR

Let us consider a convex optimization problem

$$\inf_{x \in M} \{f(x)\}! \quad , \tag{I}$$

where $M \subset E_n$ is a nonempty convex set and $f(x)$ a convex function defined over a convex set $\Omega \subset E_n$ with the property $M \subset \Omega$. This problem has the following properties:

1° The set of all its optimal solutions is convex;

2° Each local minimum of $f(x)$ over M is in the same time its global minimum.

Properties 1°, 2° have a fundamental significance both for the theoretical and for the numerical approach to convex optimization problems. In these properties consists a great advantage in comparison to other optimization problems, for which 1°, 2° do not hold.

If, e.g., S is a nonempty connected subset of a curved space $V \subset E_n$ (for instance of a smooth hypersurface) and $f(x)$ a function defined over S (or over a domain

$\Omega \subset E_n$ with the property $S \subset \Omega$), then the optimization problem

$$\inf_{x \in S} \{f(x)\} \, ! \tag{II}$$

has not in general the above mentioned properties 1° and 2°. It arises naturally a question, whether it is possible to introduce an appropriate concept of "a convex set $S \subset V$ with respect to the manifold V" and "a convex function $f(x)$ ($x \in S$) with respect to the manifold V" for a given smooth curved manifold $V \subset E_n$ in such a way that the properties 1°, 2° are satisfied for the corresponding problem (II).

Example (Steiner-Weber problem)

Unit hypersphere	$V := \{x \in E_n \mid (x, x) = 1\}$.
Open halfspace	$H := \{x \in E_n \ (a, x) < 0\} \ (a \neq o)$
Points	$x_i \in V \cap H \quad (i = 1, \ldots, N)$
Numbers	$p_i > 0 \quad (i = 1, \ldots, N)$

We define

$$S := V \cap H \, ,$$
$$f(x) := \sum_{i=1}^{N} p_i \arccos (x, x_i) \, , \quad x \in S$$

and the corresponding optimization problem

$$\inf_{x \in S} \{f(x)\} \, !$$

Remark. The concept of convexity of sets and functions can be generalized in one of the following manners so that the properties 1°, 2° are satisfied:

- pseudo- and quasiconvex functions over convex sets;
- convexity in infinite-dimensional spaces;
- convexity in curved Riemann spaces defined on the basis

of geodethical curves;

- axiomatical approach to the generalization of convexity.

Our aim consists in introducing an appropriate concept of convex set and convex function with respect to a smooth hypersurface in E_n .

1. Preliminary concepts

Let $F(\mathbf{x})$ be a function defined over the whole Euclidean space E_n $(n \geq 2)$ with the following properties:

a) continuously differentiable at least of order three in E_n , (1.1)

b) matrix $(F_{\alpha\beta})_{n\times n}$ of second partial derivatives

$$F_{\alpha\beta} = \frac{\partial^2 F}{\partial x^\alpha \partial x^\beta}$$

is positive definite in each point $\mathbf{x} = \{x^\alpha\}$,

$\mathbf{x} \in E_n \setminus A$, where

$$A := \{\mathbf{x} \in E_n \mid F_\alpha = 0 \ (\alpha = 1,\dots,n)\} \qquad (1.2)$$

$$(F_\alpha := \frac{\partial F}{\partial x^\alpha})$$

The following consequences follow from the assumed properties of $F(\mathbf{x})$:

1^0 $F(\mathbf{x})$ is strictly convex in E_n .

2^0 The set A is either empty or a one-element set.

3^0 If we define

$$\hat{\mu} := \inf_{\mathbf{x} \in E_n} \{F(\mathbf{x})\}, \qquad (1.3)$$

then for each $\mu > \hat{\mu}$ the set

$$V_\mu := \{\mathbf{x} \in E_n \mid F(\mathbf{x}) = \mu\} \tag{1.4}$$

is a smooth hypersurface (of class C^3) in the implizit sense of Monge.

4° The set

$$K_\mu := \{\mathbf{x} \in E_n \mid F(\mathbf{x}) \leq \mu\} \quad (\mu > \hat{\mu})$$

is a n-dimensional convex set in E_n with the boundary $\partial K_\mu = V_\mu$.

5° The epigraph of $F(\mathbf{x})$, i.e. the set

$$\mathcal{E}_F := \{\{\mathbf{x}, x_{n+1}\} \in E_n \times E_1 \mid x_{n+1} \geq F(\mathbf{x}), \mathbf{x} \in E_n\},$$

is a (n+1)-dimensional closed convex set in $E_{n+1} := E_n \times E_1$.

6° The spherical image S_μ of the set V_μ $(\mu > \hat{\mu})$, i.e. the set

$$S_\mu := \left\{\mathbf{z} \in E_n \mid \mathbf{z} = \frac{\nabla F(\mathbf{x})}{\|\nabla F(\mathbf{x})\|}, \mathbf{x} \in V_\mu\right\}, \tag{1.5}$$

has the following properties:

(1) $S_{\mu_1} = S_{\mu_2}$ for each pair of numbers satisfying $\mu_i > \mu$ (i = 1,2).

(2) $S_\mu \subset Q := \{\mathbf{z} \in E_n \mid \|\mathbf{z}\| = 1\}$ $(\mu > \hat{\mu})$,

where in case $A \neq \emptyset$ it is $S_\mu = Q$ and in case $A = \emptyset$ there exists an open hemisphere H_Q of the hypersphere Q (and therefore a vector $\mathbf{v} \neq \mathbf{o}$) such that

$$S_\mu \subset H_Q := \{\mathbf{z} \in Q \mid (\mathbf{z}, \mathbf{v}) > 0\}.$$

(3) The set S_μ is spherically convex, i.e. S_μ is the intersection of a convex cone with vertex $\mathbf{o}$ ($\mathbf{o}$ is origin in E_n) with the hypersphere Q .

(4) The set S_μ is relatively open in the hypersphere Q (in the sense of the usual topology in Q).

7° S_μ is a homeomorph image of manifold V_μ .

From the geometrical point of view we can say that the notion of the stright line as a privileged curve in E_n and the notion of its arc, i.e. its segment, are starting notions for the definition of convexity in E_n in classical sense. This fact leads us to the idea to introduce another generalized notion of convexity in E_n with the aid of certain privileged curves and their arcs having the following properties:

(a) The class of these curves is uniquely determined by a given function $F(\mathbf{x})$ with properties a), b) from (1.1).

(b) There exists at least one curve of this class passing through an arbitrary given pair of points $\underset{i}{\mathbf{x}} \in E_n \setminus A$ $(i = 1,2)$, $\underset{1}{\mathbf{x}} \neq \underset{2}{\mathbf{x}}$.

(c) Each curve is globally determined by one of points $\underset{0}{\mathbf{x}} \in E_n \setminus A$ and a direction $\underset{0}{\mathbf{v}}$ $(\| \underset{0}{\mathbf{v}} \| = 1)$ in this point.

(d) In case that $\underset{0}{\mathbf{v}}$ is a tangent vector of hypersurface

$$V_{\underset{0}{\mu}} := \{ \mathbf{x} \in E_n \mid F(\mathbf{x}) = \underset{0}{\mu} \} \quad (\underset{0}{\mu} := F(\underset{0}{\mathbf{x}})),$$

the corresponding curve lies in this manifold.

The curves with properties (a) - (d) will be called briefly F-curves in E_n . Further we shall introduce three classes of such F-curves, namely metrical of the first and of the second kind and affine F-curves. With the aid of these curves we shall introduce then the notion of F-convex sets and functions.

2. Metrical F-curves and metrical F-convexity of the first kind in E_n

Theorem 1. If $F(\mathfrak{x})$ has the properties (1.1) $\underset{0}{\mathfrak{x}} \in E_n \setminus A$ is a given point, $\underset{0}{\mathbf{v}}$ is a given unit vector and $\underset{0}{s}$ a given number, then there exists a one-dimensional interval I with the property $\underset{0}{s} \in \operatorname{int} I$ such that the differential equations system

$$\frac{d^2 x^\gamma}{d s^2} = \frac{\delta^{\gamma\varrho}}{\delta^{\mu\nu} F_\mu F_\nu} \left[F_\beta F_{\varrho\alpha} - F_\varrho F_{\alpha\beta} \right] \frac{dx^\alpha}{ds} \frac{dx^\beta}{ds} \qquad (2.1)_a$$

$$(\gamma = 1, \ldots, n)$$

under the initial conditions

$$x^\alpha(\underset{0}{s}) = \underset{0}{x}^\alpha, \quad \frac{dx^\alpha}{ds}(\underset{0}{s}) = \underset{0}{v}^\alpha \quad (\alpha = 1, \ldots, n) \qquad (2.1)_b$$

has a unique solution $x^\alpha(s)$ $(\alpha = 1, \ldots, n)$, $s \in I$. Further, it holds (under using the usual convention of summation from Einstein)

(1) $$\delta_{\alpha\beta}\, n^\alpha \frac{dx^\beta}{ds} = \left(\delta_{\alpha\beta}\, n^\alpha \frac{dx^\beta}{ds}\right)_{s=\underset{0}{s}} \quad \text{for all } s \in I,$$

where $$n^\alpha := \frac{\delta^{\alpha\varrho} F_\varrho}{\|\nabla F\|}.$$

(2) In case

$$\delta_{\alpha\beta}\, n^\alpha (\underset{0}{\mathfrak{x}})\, \underset{0}{v}^\beta = 0$$

is the integral curve

$$\mathcal{L} := \{ \mathfrak{x} \in E_n \mid x^\alpha = x^\alpha(s) \ (\alpha = 1, \ldots, n),\ s \in I \}$$

in the same time the geodethical curve of the hypersurface

$$V_{\mu_0} := \{ \mathbf{x} \in E_n \mid F(\mathbf{x}) = \mu_0 \} \; (\mu_0 := F(\mathbf{x}_0)) \; .$$

(3) In case

$$0 < | \delta_{\alpha\beta} \, n^{\alpha}(\mathbf{x}_0) \, v_0^{\beta} | \leqslant 1$$

$\mathcal{L}$ is an isogonal, in case

$$| \delta_{\alpha\beta} \, n^{\alpha}(\mathbf{x}_0) \, v_0^{\beta} | = 1$$

an orthogonal trajectory of the system of hypersurfaces

$$\{ V_{\mu} \}_{\mu > \hat{\mu}} \; , \text{ where } \hat{\mu} := \inf_{\mathbf{x} \in E_n} \{ F(\mathbf{x}) \} \, ,$$

for which it is $V_{\mu} \cap \mathcal{L} \neq \emptyset$.

<u>Example</u>.

In the special case $F(\mathbf{x}) = (\mathbf{x}, \mathbf{x})$ we obtain from $(1.2)_a$ the following system of differential equations:

$$\frac{d^2 x^{\gamma}}{ds^2} + \frac{x^{\gamma}}{(\mathbf{x}, \mathbf{x})} = \frac{1}{2} \left[\frac{d}{ds} \log (\mathbf{x}, \mathbf{x}) \right] \frac{dx^{\gamma}}{ds} \quad (\gamma = 1, \ldots, n) \qquad (2.2)$$

and the integral curves are either isogonal trajectories of the system of all hyperspheres

$$Q_{\mu} := \{ \mathbf{x} \in E_n \mid (\mathbf{x}, \mathbf{x}) = \mu \} \, , \; \mu > 0 \; ,$$

or the main-circles of one of them.

<u>Definition 1</u>.

We call the curve $\mathcal{L}$ defined in theorem 1 the metrical F-curve of the first kind in E_n . The arc $\mathcal{L}(\mathbf{x}_1, \mathbf{x}_2)$ joining two given points $\mathbf{x}_i \in E_n \setminus A$ $(i = 1,2)$, $\mathbf{x}_1 \neq \mathbf{x}_2$, will be called the metrical F-arc of the first kind with boundary points x_1, x_2, if the following conditions are fullfiled:

a) $\mathcal{L}(x_1, x_2)$ is an arc of a metrical F-curve of the first kind,

b) it is a shortest one of all arcs $\mathcal{L}(x_1, x_2)$ with the property a).

Definition 2.

A set $S \subset E_n$ will be called metrical F-convex of the first kind in E_n, if

(1) for each pair of points $x_i \in S$ $(i = 1,2)$, $x_1 \neq x_2$, there exists a metrical F-arc $\mathcal{L}(x_1, x_2)$ of the first kind with the property $\mathcal{L}(x_1, x_2) \subset S$,

(2) every metrical F-arc $\mathcal{L}(x_1, x_2)$ of the first kind, for which

$$S \cap \{\mathcal{L}(x_1, x_2) \setminus \{x_1, x_2\}\} \neq \emptyset$$

has the property $\mathcal{L}(x_1, x_2) \subset S$.

Examples of metrical F-convex sets (of the first kind)

The empty set, one-point set, $E_n \setminus A$, hypersurface V_μ $(\mu > \hat{\mu})$, the intersection of a hypersphere Q with the centre in o with a polyhedral cone with wertex o.

Remark.

If $F(x)$ is a linear function, $F(x) = (a, x)$ $(a \neq o)$, then $F(x)$ has not the property (1.1). In this case we obtain from $(1.2)_a$ the differential equations

$$\frac{d^2x^\gamma}{ds^2} = 0 \quad (\gamma = 1,\dots,n)$$

and the corresponding "F-convexity" is the convexity in the classical sense.

3. Metrical F-curves and metrical F-convexity of the second kind

If we define

$$h_{\alpha\beta} := \frac{F_{\alpha\beta}}{\sqrt{\delta^{\mu\nu}F_\mu F_\nu}} \qquad (\alpha,\beta = 1,\dots,n),$$

then - under the assumption about the function $F(\boldsymbol{x})$ given above - the matrix $(h_{\alpha\beta})_{n\times n}$ is positive definit in each point $\underset{0}{\boldsymbol{x}} \in E_n \setminus A$. If we consider, that $h_{\alpha\beta}$ represent a tensor over $E_n \setminus A$ and if we define the connection in the following way

$$\left\{ {\gamma \atop \alpha\beta} \right\} := \frac{1}{2} h^{\gamma\rho}(\partial_\alpha h_{\rho\beta} + \partial_\beta h_{\alpha\rho} - \partial_\rho h_{\alpha\beta}), \qquad (3.1)_a$$

where $h^{\alpha\beta}$ is the counter-gradient tensor to the tensor $h_{\alpha\beta}$, then we can form the following system of differential equations

$$\frac{d^2x^\gamma}{ds^2} + \left\{ {\gamma \atop \alpha\beta} \right\} \frac{dx^\alpha}{ds}\frac{dx^\beta}{ds} = 0 \quad (\gamma = 1,\dots,n). \qquad (3.1)_b$$

If $\underset{0}{\boldsymbol{x}} \in E_n \setminus A$ is a given point, s_0 a given number and $\underset{0}{\boldsymbol{v}} \neq \boldsymbol{o}$ a given vector, then there exists exactly one solution curve $\mathcal{L}$ of this system $(1.3)_b$ with the property $\underset{0}{\boldsymbol{x}} \in \mathcal{L}$ and with the tangential vector $\underset{0}{\boldsymbol{v}}$ in $\underset{0}{\boldsymbol{x}}$. If

$$\mathcal{L} := \{\boldsymbol{x} \in E_n \setminus A \mid x^\alpha = x^\alpha(s) \ (\alpha = 1,\dots,n),\ s \in I\}$$

is the corresponding solution curve with the property $s_0 \in \operatorname{int} I$, then it is

$$x^{\alpha}(\underset{0}{s}) = \underset{0}{x}^{\alpha}, \quad \frac{dx^{\alpha}}{ds}(\underset{0}{s}) = \underset{0}{v}^{\alpha} \quad (\alpha = 1,\dots,n).$$

In the case $\underset{0}{\mathbf{v}} \neq \mathbf{o}$, $(\underset{0}{\mathbf{v}}, \nabla F(\underset{0}{\mathbf{x}})) = 0$, the corresponding curve $\mathcal{L}$ is contained in the hypersurface $V_{\underset{0}{\mu}}$ with $\underset{0}{\mu} = F(\underset{0}{\mathbf{x}})$.

The solution curves of $(1.3)_b$ are briefly called metrical F-curves of the second kind. On the basis of such curves we can introduce the notion of the metrical F-convexity of the second kind similary as in the case of the convexity of the first kind.

4. Affine F-curves and affine F-convexity in E_n

Let us consider the hypersurface

$$V_{\mu} = \{\mathbf{x} \in E_n \mid F(\mathbf{x}) = \mu\},$$

where μ is a given number with $\mu > \hat{\mu}$ and $\underset{1}{\mathbf{v}}, \underset{2}{\mathbf{v}}$ are given orthonormal vectors. If we define

$$\mathbf{n}(\varphi) := \underset{1}{\mathbf{v}} \cos\varphi + \underset{2}{\mathbf{v}} \sin\varphi, \quad \varphi \in \langle 0, 2\pi),$$

then the optimization problem

$$\max_{\mathbf{x} \in V_{\mu}} \{(\mathbf{n}(\varphi), \mathbf{x})\} \; ! \tag{4.1}$$

has in case $A \neq \emptyset$ a unique optimal point $\mathbf{x}(\varphi)$ for every $\varphi \in \langle 0, 2\pi)$ and in case $A = \emptyset$ there exists at most one optimal point for every $\varphi \in \langle 0, 2\pi)$. However, in both cases, there exists some maximal interval $I \subset \langle 0, 2\pi)$ of parameter φ representing the solubility domain of the problem (1.3) and therefore

$$(\mathbf{n}(\varphi), \mathbf{x}(\varphi)) = \max_{\mathbf{x} \in V_\mu} \{(\mathbf{n}(\varphi), \mathbf{x})\}, \varphi \in I.$$

<u>Theorem 2</u>. The set

$$\mathcal{L} := \{\mathbf{x} \in E_n \mid \mathbf{x} = \mathbf{x}(\varphi), \varphi \in I\}$$

is a smooth curve in E_n with the property $\mathcal{L} \subset V_\mu$, and the equations

$$(\nabla F(\mathbf{x}), \underset{i}{\mathbf{v}}) = 0 \quad (i \in \{1,\dots,n\} \setminus \{1,2\})$$

$$F(\mathbf{x}) = \mu,$$

where $\underset{\alpha}{\mathbf{v}}$ $(\alpha = 1,\dots,n)$ are vectors with the property $(\underset{\alpha}{\mathbf{v}}, \underset{\beta}{\mathbf{v}}) = \delta_{\alpha\beta}$ $(\alpha, \beta = 1,\dots,n)$, denote an implicit description of $\mathcal{L}$.

<u>Definition 3</u>.

The curve $\mathcal{L}$ from theorem 2 will be called affine geodetical curve of the manifold V_μ.

Let us consider now the whole system $\{V_\mu\}_{\mu > \hat{\mu}}$ of hypersurfaces. Let $\underset{0}{\mathbf{x}} \in E_n \setminus A$ and $\underset{i}{\mathbf{v}}$ $(i = 1,2)$ are two orthonormal vectors. If we define ($w \geq 0$ a given number)

$$\mathbf{n}(\mu) := \underset{1}{\mathbf{v}} \cos w(\mu - \mu_0) + \underset{2}{\mathbf{v}} \sin w(\mu - \mu_0), \ \mu > \hat{\mu},$$

where $\mu_0 := F(\underset{0}{\mathbf{x}})$, then the parametric optimization problem

$$\max_{\mathbf{x} \in V_\mu} \{(\mathbf{n}(\mu), \mathbf{x})\}!, \ \mu > \hat{\mu} \tag{4.2}$$

has in case $A \neq \emptyset$ a unique optimal point $\mathbf{x}_w(\mu)$ for each $\mu > \hat{\mu}$ and in the case $A = \emptyset$ at most one for every $\mu \in (\hat{\mu}, \infty)$. If we define

$$\mathbf{v}_1 := \frac{\nabla F(\mathbf{x}_0)}{\|\nabla F(\mathbf{x}_0)\|} \; ,$$

then the problem (2.3) is soluble and there exists some maximal interval I_0 with $\mu_0 \in \operatorname{int} I_0$ in such a way, that the problem (2.3) has for every $\mu \in I_0$ exactly one optimal point $\mathbf{x}_w(\mu)$.

Theorem 3. The set

$$\mathscr{L}_w := \left\{ \mathbf{x} \in E_n \mid \mathbf{x} = \mathbf{x}_w(\mu), \mu \in I_0 \right\} \tag{4.3}$$

is a smooth curve in E_n with the property $\mathbf{x}_0 \in \mathscr{L}_w$. This curve is a certain connected component of the curve (in the sense of Monge) with the implicit description:

$$(\nabla F(\mathbf{x}), \mathbf{v}_i) = 0 \quad (i \in \{1,\dots,n\} \setminus \{1,2\}),$$

$$k \arccos \left(\frac{\nabla F(\mathbf{x})}{\|\nabla F(\mathbf{x})\|}, \mathbf{v}_1 \right) = \sqrt{1-k^2} \, (F(\mathbf{x}) - F(\mathbf{x}_0)),$$

$$k \arcsin \left(\frac{\nabla F(\mathbf{x})}{\|\nabla F(\mathbf{x})\|}, \mathbf{v}_2 \right) = \sqrt{1-k^2} \, (F(\mathbf{x}) - F(\mathbf{x}_0)),$$

where

$$k := \frac{1}{\sqrt{1+w^2}} \; , \quad (\mathbf{v}_\alpha, \mathbf{v}_\beta) = \delta_{\alpha\beta} \quad (\alpha, \beta = 1,\dots,n).$$

The image of the curve $\mathscr{L}_w$ under the regular simple mapping

$$\mathbf{z} = e^{F(\mathbf{x})} \frac{\nabla F(\mathbf{x})}{\|\nabla F(\mathbf{x})\|} \; , \quad \mathbf{x} \in E_n \setminus A$$

is a metrical F-curve for $F(\mathbf{z}) = (\mathbf{z}, \mathbf{z})$, or its connected subset; it is therefore a connected part of an isogonal trajectory of the system of hyperspheres with centre in the origin $\mathbf{o}$. The image of an affine geodethical curve of the manifold V_μ is then a main-circle of a certain hypersphere of this system (or certain arc of it).

Remark. If $\mathbf{v} \neq \mathbf{o}$ is an arbitrary vector and

$$P(\mathbf{v}) := \{ \mathbf{x} \in E_n \mid (\nabla F(\mathbf{x}), \mathbf{v}) = 0 \}$$

is not empty, then $P(\mathbf{v})$ is a regular hypersurface in E_n separating E_n in two domains

$$P^+(\mathbf{v}) := \{ \mathbf{x} \in E_n \mid (\nabla F(\mathbf{x}), \mathbf{v}) > 0 , \quad P^-(\mathbf{v}) := \{ \mathbf{x} \in E_n \mid (\nabla F(\mathbf{x}), \mathbf{v}) < 0 \}$$

and for each pair of points $\mathbf{x}_i \in E_n$ $(i = 1,2)$ there exists a vector $\mathbf{v} \neq \mathbf{o}$ such that $\mathbf{x}_i \in P^+(\mathbf{v})$ $(i = 1,2)$.

The set $P^+(\mathbf{v})$ (or $P^-(\mathbf{v})$) will be called open affine F-halfspace in E_n.

Definition 4.

The curve $\mathcal{L}_w$ from (3.3) or the affine geodethical curve of the hypersurface V_μ will be called affine F-curve in E_n. An arc $\mathcal{L}(\mathbf{x}_1, \mathbf{x}_2)$ joining two points $\mathbf{x}_i \in E_n$ $(i = 1,2)$, $\mathbf{x}_1 \neq \mathbf{x}_2$, will be called the affine F-arc with the boundary points $\mathbf{x}_1, \mathbf{x}_2$, if

a) $\mathcal{L}(\mathbf{x}_1, \mathbf{x}_2)$ is an arc of an affine F-curve,

b) there exists a vector $\mathbf{v} \neq \mathbf{o}$ with the property

$$\mathcal{L}(\mathbf{x}_1, \mathbf{x}_2) \subset \overline{P}^+(\mathbf{v}).$$

Definition 5.

We call a set $S \subset E_n$ an affine F-convex set in E_n if

(1) for each pair $\mathbf{x}_i \in S$ $(i = 1,2)$, $\mathbf{x}_1 \neq \mathbf{x}_2$, there exists an affine F-arc $\mathcal{L}(\mathbf{x}_1, \mathbf{x}_2)$ (with boundary points $\mathbf{x}_1$, $\mathbf{x}_2$) with the property $\mathcal{L}(\mathbf{x}_1, \mathbf{x}_2) \subset S$;

(2) every affine F-arc $\mathcal{L}(\mathbf{x}_1, \mathbf{x}_2)$, for which

$$S \cap \{\mathcal{L}(\mathbf{x}_1, \mathbf{x}_2) \setminus \{\mathbf{x}_1, \mathbf{x}_2\}\} \neq \emptyset,$$

has the property $\mathcal{L}(\mathbf{x}_1, \mathbf{x}_2) \subset S$.

Final Remark. The notion of F-convex function $f(\mathbf{x})$ over a F-convex set $S \subset E_n$ - either in the metrical or in the affine sense - can be introduced with the aid of the inequality

$$f(\mathbf{x}(\lambda_1 s_1 + \lambda_2 s_2)) \leq \lambda_1 f(\mathbf{x}_1) + \lambda_2 f(\mathbf{x}_2), \quad \mathbf{x}_i \in S \ (i = 1,2), \ \lambda_i \geq 0 \ (i = 1,2), \ \lambda_1 + \lambda_2 = 1,$$

where the function $\mathbf{x}(s)$ is the function used in the description of the corresponding F-curve and $\mathbf{x}_i = \mathbf{x}(s_i)$ $(i = 1,2)$. The parameter s denotes here either the metrical arc (in the case of metrical F-curve) or the affine arc (in the case of affine F-curve).

Prof.Dr. František Nožička, Faculty of Mathematics and Physics, Charles University, Praha 1, Malostranské nám. 25, ČSSR

International Series of
Numerical Mathematics, Vol. 72

UNICITY IN SEMI-INFINITE OPTIMIZATION

Günther Nürnberger

Fakultät für Mathematik und Informatik, Universität Mannheim, 6800 Mannheim, Federal Republic of Germany

Abstract

In practice the parameter which determines a semi-infinite optimization problem is only known approximatively or changes. A complete characterization of linear optimization problems which have a strongly unique solution and are stable under small changes of the parameter is given. Moreover, the relationship between unique and strongly unique solutions is examined.

Introduction

The unicity of solutions for the following linear optimization problem is examined. Let $p \in \mathbb{R}^N$, $B \in C(T, \mathbb{R}^N)$, $b \in C(T)$ and $\sigma=(p,B,b)$ be given. Minimize

$$\langle p,x\rangle = \sum_{i=1}^{N} p_i x_i \text{ subject to } \langle B(t),x\rangle \leqq b(t) \text{ for all } t \in T.$$

This problem is denoted by $LM(\sigma)$. All considered problems shall satisfy the so-called Slater-condition, i.e. that there exists a vector $y \in \mathbb{R}^N$ such that $\langle B(t),y\rangle < b(t)$ for all $t \in T$.

In practice it is natural to consider the following type of unicity (compare Example 2.1). A solution $x \in \mathbb{R}^N$ of $LM(\sigma)$

is said to be strongly unique, if there exists a constant $K>0$ such that for all feasible points $y \in \mathbb{R}^N$,

$$\langle p,y\rangle \geq \langle p,x\rangle + K\,\|x-y\|_2 .$$

We denote by SU the set $SU = \{\sigma: LM(\sigma)$ has a strongly unique solution$\}$.

Characterizations of strong unicity in optimization and best approximation can be found in Bartelt & McLaughlin [1], Brosowski [3], [4], Hettich & Zencke [7], Nürnberger, Schumaker, Sommer & Strauß [14], Wulbert [15] and in [10], [11]. Since in practice the parameter σ may be only known approximatively, this leads to the question which $\sigma \in SU$ are "stable under small changes", i.e. $\sigma \in$ interior SU.

In section 1 we give a complete characterization of such parameters. Moreover, we consider similar questions for the case when e.g. only p or b is variable. The results can be applied to best approximation and yield a characterization given in [12], where in particular spline approximation is considered.

In section 2 we investigate the relationship between unique and strongly unique solutions. It is well-known that they are not the same in general. However, it is shown that in finite optimization (i.e. T finite) unique and strongly unique solutions coincide. Furthermore, a density result is given. It says that if LM(p,B,b) has a unique solution, then b can be "changed slightly" such that the corresponding problem has a strongly unique solution. Results of this type have been given in Nürnberger & Singer [13] for best approximation.

1. Strong unicity

For a compact metric space T we denote by C(T) (resp. $C(T, \mathbb{R}^N)$) the set of all continuous mappings $b: T \to \mathbb{R}$ (resp. $B: T \to \mathbb{R}^N$) endowed with the norm
$\|b\|_\infty = \sup\{|b(t)|: t \in T\}$ (resp. $\|B\|_\infty = \sup\{\|B(t)\|_2 : t \in T\}$).
Let $p \in \mathbb{R}^N$, $B \in C(T, \mathbb{R}^N)$, $b \in C(T)$ and $\sigma=(p,B,b)$ be given.

We consider the following <u>linear optimization problem</u> LM(σ).

$$\text{Minimize } \langle p,x\rangle = \sum_{i=1}^{N} p_i x_i \text{ subject to}$$

$$\langle B(t),x\rangle \leq b(t) \text{ for all } t \in T.$$

Moreover, we set $\|\sigma\| = \|p\|_2 + \|B\|_\infty + \|b\|_\infty$.

The set of <u>feasible points</u> is defined by

$$Z_\sigma = \{x \in \mathbb{R}^N : \langle B(t),x\rangle \leq b(t) \text{ for all } t \in T\},$$

the <u>minimum value</u> by

$$E_\sigma = \inf \{\langle p,x\rangle : x \in Z_\sigma\}$$

and the set of <u>solutions</u> by

$$P_\sigma = \{x \in Z_\sigma : \langle p,x\rangle = E_\sigma\}.$$

For a given vector $x \in Z_\sigma$ the set of <u>active points</u> is defined by

$$N_{\sigma,x} = \{t \in T : \langle B(t),x\rangle = b(t)\}.$$

Let t_o be any point not in T. We set $B(t_o) = p$ and $b(t_o) = E_\sigma$. A closed subset M of T is called <u>critical</u> (with respect to (p,B)), $\quad 0 \in \operatorname{conv} \{B(t) \in \mathbb{R}^N : t \in M \cup \{t_o\}\}$.

(Here conv(.) denotes the convex hull of a set.)

It is well-known (see Brosowski [4]) that the last condition is equivalent to

$$\min \{\langle B(t),y\rangle : t \in M \cup \{t_o\}\} \leq 0 \quad \text{for all } y \in \mathbb{R}^N.$$

The optimization problem LM(σ) is said to satisfy the <u>Slater-condition</u>, if there exists a vector $y \in \mathbb{R}^N$ such that

$$\langle B(t),y\rangle < b(t) \quad \text{for all } t \in T.$$

We denote by Q the set $Q = \{\sigma : \text{LM}(\sigma) \text{ satisfies the Slater-condition}\}$ and by L the set $L = \{\sigma \in Q : \text{LM}(\sigma) \text{ has a solution}\}$.

Concerning uniqueness of solutions, in practice it is natural to require the following property. A solution $x \in \mathbb{R}^N$ of LM(σ) is called <u>strongly unique</u>, if there exists a constant $K>0$ such that for all $y \in Z_\sigma$,

$$\langle p,y\rangle \geq \langle p,x\rangle + K \|x-y\|_2 .$$

We denote by SU the set $SU = \{\sigma \in Q : \text{LM}(\sigma) \text{ has a strongly unique solution}\}$ and by U the set $U = \{\sigma \in Q : \text{LM}(\sigma) \text{ has a unique solution}\}$.

We first recall a famous characterization of solutions (see e.g. Brosowski [4,p.62]).

Theorem 1.1 (Kuhn-Tucker) For $\sigma \in Q$ and $x \in Z_\sigma$ the following conditions are equivalent:
(1) x is a solution of LM(σ).
(2) There exists a critical subset $\{t_1,\ldots,t_q\}$ of $N_{\sigma,x}$ $(q \leq N)$.

Moreover, we need the following results on strong unicity (see Hettich & Zencke [7, p. 70]).

Theorem 1.2 For $\sigma \in Q$ and $x \in P_\sigma$ the following conditions are equivalent:
(1) $\sigma \in SU$.
(2) For each $y \in \mathbb{R}^N$, $y \neq 0$, there exists a point $t \in N_{\sigma,x} \cup \{t_0\}$ with $\langle B(t),y \rangle < 0$.

The next result is a consequence of Theorem 1.2.

Theorem 1.3 For $\sigma \in Q$ and $x \in P_\sigma$ we consider the following conditions:
(1) $\sigma \in SU$.
(2) There exists a critical subset $\{t_1,\ldots,t_N\}$ of $N_{\sigma,x}$ such that

$$B(t_0), B(t_1), \ldots, B(t_{i-1}), B(t_{i+1}), \ldots, B(t_N)$$

are linearly independent, $i=0,1,\ldots,N$.
Then (2) $\Rightarrow$ (1) and, if card $N_{\sigma,x} = N$, then (1) $\Rightarrow$ (2).
Moreover, if $\sigma \in SU$, then there exists a critical subset of $N_{\sigma,x}$ consisting of N distinct points.

Unfortunately, condition (1) and (2) in Theorem 1.3 are not equivalent, even in the particular case of best approximation. Further characterizations of strong unicity have been given in [1], [3], [4], [7], [10], [11], [14] and [15].

In practice it is natural to ask which parameters $\sigma \in SU$ are "stable under small changes", i.e. which σ are from int SU. (Here int denotes the interior of a set.) We give the following characterization.

Theorem 1.4 For $\sigma \in Q$ and $x \in P_\sigma$ the following conditions (1) and (2) are equivalent.

(1) $\sigma \in \text{int } SU$.

(2) (a) There exists a critical subset $\{t_1,\ldots,t_N\}$ of $N_{\sigma,x}$ consisting of N distinct points.

(b) For every such set $\{t_1,\ldots,t_N\}$

$$B(t_o),B(t_1),\ldots,B(t_{i-1}),B(t_{i+1}),\ldots,B(t_N)$$

are linearly independent, $i=0,1,\ldots,N$.

For the proof of this theorem we need the following two lemmas.

Lemma 1.5 If $\sigma \in L$ and P_σ is compact, then there exists a real number $\varepsilon>0$ such that for all parameters $\tilde{\sigma}$ with $\|\sigma-\tilde{\sigma}\|<\varepsilon$ we have $\tilde{\sigma} \in L$.

Proof. Let $\sigma \in L$ be given such that P_σ is compact. By compactness arguments there exists an $\varepsilon_1>0$ such that for all $\tilde{\sigma}$ with $\|\sigma-\tilde{\sigma}\|<\varepsilon_1$ we have $\tilde{\sigma} \in Q$. Now, we assume that there exists a sequence $(\sigma_n) \subset Q$ such that $\sigma_n \to \sigma$ and $P_{\sigma_n}=\phi$ for all n.

Let an integer n be given. Since $Z_{\sigma_n} \neq \phi$, there exists a sequence $(y_m^{(n)}) \subset Z_{\sigma_n}$ with $\lim_{m\to\infty} \langle p_n,y_m^{(n)}\rangle = E_{\sigma_n}$. Since $P_{\sigma_n}=\phi$, the sequence $(y_m^{(n)})$ is unbounded. This shows that we may choose an unbounded sequence (y_n) such that for all n,

(1) $y_n \in Z_{\sigma_n}$.

(2) $\langle p_n,y_n\rangle \leq E_{\sigma_n}+\frac{1}{n}$, if $E_{\sigma_n}>-\infty$.

(3) $\langle p_n,y_n\rangle \leq 0$, if $E_{\sigma_n}=-\infty$.

By passing to a subsequence we may assume that $\|y_n\|_2\to\infty$ and $y_n/\|y_n\|_2 \to y \in \mathbb{R}^N$. Now, we choose a point $w \in Z_\sigma$. By Theorem 12, (2a) on p. 216 in Brosowski [4] there exists a sequence (w_n)

such that $w_n \in Z_{\sigma_n}$ and $w_n \to w$. Then by (2) there exists a constant $K_1 > 0$ such that for all n with $E_{\sigma_n} > -\infty$,

$\langle p_n, y_n \rangle \leqq E_{\sigma_n} + \frac{1}{n} \leqq \langle p_n, w_n \rangle + \frac{1}{n} \leqq \langle p, w \rangle + K_1 + 1.$

Then it follows from (3) that there exists a constant $K_2 > 0$ such that $\langle p_n, y_n \rangle \leqq K_2$ for all n. Therefore

$\langle p_n, y_n / \|y_n\|_2 \rangle \leqq K_2 / \|y_n\|_2$ for all n.

By taking limits, it follows that

(4) $\langle p, y \rangle \leqq 0$.

Furthermore, it follows from (1) that

$\langle B_n(t), y_n / \|y_n\|_2 \rangle \leqq b_n(t) / \|y_n\|_2$ for all $t \in T$.

By taking limits, it follows that

(5) $\langle B(t), y \rangle \leqq 0$ for all $t \in T$.

Now, we choose a point $v \in P_\sigma$. Then it follows from (4) and (5) that $v + \lambda y \in P_\sigma$ for all $\lambda \geqq 0$. This contradicts the compactness of P_σ and proves Lemma 1.5.

<u>Lemma 1.6</u> If $\sigma \in L$ and P_σ is compact, then for each sequence (σ_n) in L with $\sigma_n \to \sigma$ and for each sequence (x_n) with $x_n \in P_{\sigma_n}$ there exists a subsequence of (x_n) converging to some point $x \in P_\sigma$.

<u>Proof</u>. Let $\sigma_n \to \sigma$ and $x_n \in P_{\sigma_n}$ for all n be given. If (x_n) would be unbounded, then by the proof of Lemma 1.5 we would get a contradiction to the compactness of P_σ. Thus going to a subsequence we may assume that (x_n) converges to some $x \in \mathbb{R}^N$. Then it follows that $x \in Z_\sigma$.

Now, we choose $y \in P_\sigma$. By Theorem 12, (2a) on p. 216 in Brosowski [4] there exists a sequence (y_n) with $y_n \in Z_{\sigma_n}$ such that $y_n \to y$. Therefore $\langle p_n, x_n \rangle \leqq \langle p_n, y_n \rangle$ for all n. Taking limits it follows that $\langle p, x \rangle \leqq \langle p, y \rangle$, which implies that $x \in P_\sigma$. This proves Lemma 1.6.

Now we are in position to give a proof of Theorem 1.4.

Proof of Theorem 1.4. We first show that (2) $\Rightarrow$ (1). We assume that (2) holds but (1) fails. It follows from Theorem 1.3 that $\sigma \in SU$. By Lemma 1.5 there exists a sequence (σ_n) in L such that $\sigma_n \to \sigma$ and $LM(\sigma_n)$ has a solution x_n which is not strongly unique. Since x_n is a solution of $LM(\sigma_n)$, it follows from Theorem 1.1 that there exists an integer $q_n \in \{1,\dots,N\}$ and a subset $M_n = \{t_{1,n},\dots,t_{q_n,n}\}$ of N_{σ_n,x_n} such that for all $y \in \mathbb{R}^N$,

(3) $\min \{\langle B(t),y\rangle : t \in M_n \cup \{t_0\}\} \leq 0$.

Since $\sigma_n \to \sigma$, $x_n \in P_{\sigma_n}$ for all n and $P_\sigma = \{x\}$, it follows from Lemma 1.6 that $x_n \to x$.

Let for all n, $\sigma_n = (p_n, B_n, b_n)$. Since $\|\sigma - \sigma_n\| \to 0$, it follows that $\|p-p_n\|_2 \to 0$, $\|B-B_n\|_\infty \to 0$ and $\|b-b_n\|_\infty \to 0$. Going to a subsequence we may assume that $q_n = q \in \{1,\dots,N\}$ for all n and that $t_{i,n} \to t_i \in T$, $i=1,\dots,q$. Since M_n is a subset of N_{σ_n,x_n}, it follows that $\langle B(t_{i,n}),x_n\rangle = b_n(t_{i,n})$, $i=1,\dots,q$. Taking limits we get $\langle B(t_i),x\rangle = b(t_i)$, $i=1,\dots,q$, which implies that $M=\{t_1,\dots,t_q\}$ is a subset of $N_{\sigma,x}$.

Let $y \in \mathbb{R}^N$ be given. It follows from (3) that for all n there exists an integer $j_n \in \{0,1,\dots,q\}$ such that $\langle B(t_{j_n,n}),y\rangle \leq 0$.

Taking limits this shows that M is critical. By omitting some points of M, if necessary, we may assume that the points of M are distinct.

Case 1. $q \in \{1,\dots,N-1\}$.

By (2a) we can choose points $t_{q+1},\dots,t_N$ in $N_{\sigma,x}$ such that all points of $\{t_1,\dots,t_N\}$ are distinct. Since $\{t_1,\dots,t_N\}$ is a critical subset of $N_{\sigma,x}$, it follows from (2b) that

(4) $B(t_0), B(t_1),\dots,B(t_{N-1})$ are linearly independent.

We consider the following system of linear equations.

$$\sum_{j=1}^{N} y_j B_j(t_i) = 1 \ , \ i=0,1,\dots,N-1.$$

where $B(t) = (B_1(t),\dots,B_N(t))$.

By (4) the corresponding determinant is different from zero, which implies that the system has a unique solution

$y=(y_1,\dots,y_N)$. Thus for all $i\in\{0,1,\dots,N-1\}$, $\langle B(t_i),y\rangle > 0$, which contradicts the fact that $\{t_1,\dots,t_q\}$ is a critical subset of $N_{\sigma,x}$.

Case 2. $q=N$.

In this case $M_n=\{t_{1,n},\dots,t_{N,n}\}$ for all n. Assume that there exists an integer n such that

$B(t_{0,n}),\dots,B(t_{i-1,n}),B(t_{i+1,n}),\dots,B(t_{N,n})$ are linearly independent, $i=0,1,\dots,N$. Then, since M_n is a critical subset of N_{σ_n,x_n}, it follows from Theorem 1.3 that $\sigma_n\in SU$, a contradiction.

Thus going to a subsequence we may assume that there exists an integer $j\in\{1,\dots,N\}$ such that for all n,

$B(t_{0,n}),B(t_{1,n}),\dots,B(t_{j-1,n}),B(t_{j+1,n}),\dots,B(t_{N,n})$

are linearly dependent. By taking limits it follows that

$B(t_0),B(t_1),\dots,B(t_{j-1}),B(t_{j+1}),\dots,B(t_N)$

are linearly dependent. Since $\{t_1,\dots,t_n\}$ is a critical subset of $N_{\sigma,x}$ this contradicts (2b). This shows that (2) $\Rightarrow$ (1).

Now we show that (1) $\Rightarrow$ (2). It follows from Theorem 1.2 that (1) $\Rightarrow$ (2a). We show that (1) $\Rightarrow$ (2b).

We assume that (2b) fails, i.e. that there exists a critical subset $\{t_1,\dots,t_N\}$ of $N_{\sigma,x}$ and an integer $j\in\{0,1,\dots,N\}$ such that

(5) $B(t_0),B(t_1),\dots,B(t_{j-1}),B(t_{j+1}),\dots,B(t_N)$ are linearly dependent.

We show that there exists a sequence (σ_n) such that $\sigma_n\to\sigma$ and for all n, $x\in Z_{\sigma_n}$ and $N_{\sigma_n,x}=\{t_1,\dots,t_N\}$.

For each $i\in\{1,\dots,N\}$ we choose an open neighborhood V_i of t_i such that the neighborhoods are disjoint. By Urysohn's lemma for each n there exists a function $h_n\in C(T)$ such that

$h_n(t)=0$ for all $t\in\{t_1,\dots,t_N\}$,

$h_n(t)=1/n$ for all $t\in T\setminus \bigcup_{i=1}^{N} V_i$ and

$0<h_n(t)<1/n$ for all $t\in \bigcup_{i=1}^{N} V_i$.

For all n we set $b_n = b + h_n$ and $\sigma_n = (p,B,b_n)$. Then obviously the sequence (σ_n) has the desired property.
Since $\sigma \in Q$, we may assume that (σ_n) in Q. Therefore, since $N_{\sigma_n,x}$ is a critical set, it follows from Theorem 1.1 that x is a solution of $LM(\sigma_n)$. But it follows from (5) and Theorem 1.3 that x is not a strongly unique solution of $LM(\sigma_n)$. Therefore $\sigma_n \notin SU$. This shows that $\sigma \notin$ int SU and proves (1) $\Rightarrow$ (2).
This completes the proof of Theorem 1.4.

Using Theorem 1.3 and Theorem 1.4 we immediately obtain the following result.

Corollary 1.7 If $\sigma \in Q$, $x \in P_\sigma$ and card $N_{\sigma,x} = N$, then $\sigma \in SU$ if and only if $\sigma \in$ int SU.

In finite optimization (i.e. T finite) Corollary 1.7 says that, if $x \in P_\sigma$ is a non-degenerate corner, then $\sigma \in SU$ if and only if $\sigma \in$ int SU.

Now, we give a simple example which shows that there exist parameters $\sigma \in SU \setminus \text{int } SU$. (Moreover, general results on spline approximation concerning such type of parameters are given below.)

Example 1.8 We consider the following optimization problem $LM(\sigma)$. Let real numbers $c_1 < 0$ and $c_2 < 0$ be given.

Minimize $\langle p,x\rangle = x_1 + x_2$ subject to

$x_1 + x_2 \leqq 1$, $-x_1 \leqq 0$, $-x_2 \leqq 0$ and $c_1x_1 + c_2x_2 \leqq 0$.

Using Theorem 1.3 and Theorem 1.4 it can be easily verified that $x=(0,0) \in P_\sigma$ and $\sigma \in SU$. But $\sigma \in$ int SU if and only if $c_1 \neq c_2$.

Now, we consider "stability questions" for the case when e.g. only p or b is variable. For fixed p and B we set $SU_{p,B} = \{b: (p,B,b) \in SU\}$.

Remark 1.9 (1) The arguments in the proof of Theorem

1.4 can be used to obtain the following results. Theorem 1.4 remains true for the case when only b is variable (i.e. for fixed p and B we have $b \in \text{int } SU_{p,B}$ if and only if condition (2) in Theorem 1.4 holds). The same is true, if only (p,b) or (B,b) is variable.

(2) Obviously condition (2) in Theorem 1.4 always provides a sufficient condition for "stability", if all parameters p,B and b or some of them are variable. However, e.g. if only B is variable, then in general this condition is not necessary.

Now, we consider the case when only p is variable. In contrast to Theorem 1.4 the following result holds. For fixed B and p we set $SU_{B,b} = \{p: (p,B,b) \in SU\}$.

<u>Theorem 1.10</u> Let fixed mappings $B \in C(T, \mathbb{R}^N)$ and $b \in C(T)$ be given. Then $SU_{B,b}$ is an open set.

<u>Proof</u>. Let $p \in SU_{B,b}$ be given and $\sigma=(p,B,b)$. Then $\sigma \in Q$ and $LM(\sigma)$ has a strongly unique solution $x \in \mathbb{R}^N$. Theorem 1.2 implies that for all $y \in \mathbb{R}^N$, $y \neq 0$,

$$\min\{\langle B(t),y\rangle: t \in N_{\sigma,x} \cup \{t_0\}\} < 0.$$

It follows from compactness arguments that there exists a constant K>0 such that for all $y \in \mathbb{R}^N$,

$$\min\{\langle B(t),y\rangle: t \in N_{\sigma,x} \cup \{t_0\}\} \leq -K \|y\|_2.$$

Let $\tilde{p} \in \mathbb{R}^N$ with $\|p-\tilde{p}\|_2 < K$ and $y \in \mathbb{R}^N$ be given. Then there exists a point $t_y \in N_{\sigma,x}$ with $\langle B(t_y),y\rangle \leq -K \|y\|_2$ or $\langle p,y\rangle \leq -K \|y\|_2$.

In the second case

$$\langle \tilde{p},y\rangle = \langle p,y\rangle + \langle \tilde{p}-p,y\rangle \leq \langle p,y\rangle + \|\tilde{p}-p\|_2 \|y\|_2$$
$$< -K \|y\|_2 + K \|y\|_2 = 0.$$

Then it follows from Theorem 1.2 that x is a strongly unique solution of $LM(\tilde{p},B,b)$, which implies that $\tilde{p} \in SU_{B,b}$. This proves Theorem 1.10.

It is well-known that best approximation is a special case of semi-infinite optimization (see any of the books in the references). The analog of Theorem 1.4 for the case when only b is variable (see Remark 1.9) yields a complete characterization of "stability" of functions having a strongly unique best approximation from arbitrary finite dimensional subspaces of C(T), given in Theorem 1.1 of [12]. (One should only observe that the functions $b \in C(T)$ for the case of best approximation are of special type. A slight modification of the proof of Theorem 1.4 also works in this case.) Instead of stating this general result to illustrate Theorem 1.3 we give a particular application to spline approximation from [12].

We denote by $S_{n,k}$ the subspace of spline functions of degree n with k fixed knots $a=x_o<x_1<\ldots<x_k<x_{k+1}=b$ in a given interval $[a,b]$. A spline function $s_f \in S_{n,k}$ is called <u>strongly unique best approximation</u> of a given function $f \in C[a,b]$, if there exists a constant $K>0$ such that for all $s \in S_{n,k}$,

$$\|f-s\|_\infty \geq \|f-s_f\|_\infty + K\|s-s_f\|_\infty .$$

We set $SU(S_{n,k}) = \{f \in C[a,b]: f$ has a strongly unique best approximation from $S_{n,k}\}$. Points $a \leq t_1<\ldots<t_q \leq b$ are called <u>alternating extreme points</u> of a given function $h \in C[a,b]$, if there exists a sign $\lambda \in \{-1,1\}$ such that

$$\lambda(-1)^i h(t_i) = \|h\|_\infty , \quad i=1,\ldots,q.$$

<u>Theorem 1.11</u> Let $s_f \in S_{n,k}$ be a best approximation of $f \in C[a,b] \setminus S_{n,k}$. Then the following statements (1) and (2) are equivalent.

(1) $f \in \operatorname{int} SU(S_{n,k})$.

(2) (a) $f-s_f$ has at least $n+k+2$ alternating extreme points.

(b) $f-s_f$ has at most $n+q$ alternating extreme points in each knot-interval $[x_p,x_{p+q}] \subset [a,b]$, $[x_p,x_{p+q}] \neq [a,b]$.

It may be of interest to compare Theorem 1.11 with the following characterization of strongly unique best spline approximations, given in [10], [11].

Theorem 1.12 Let $s_f \in S_{n,k}$ be a best approximation of $f \in C[a,b] \setminus S_{n,k}$. Then the following statements (1) and (2) are equivalent.

(1) $f \in SU(S_{n,k})$.

(2) (a) $f-s_f$ has at least $n+k+2$ alternating extreme points.

(b) $f-s_f$ has at least $j+1$ alternating extreme points in each knot-interval $[x_o,x_j)$, $(x_{k+1-j},x_{k+1}]$, $(x_i,x_{i+j+n}) \subset [a,b]$ $(j \geq 1)$.

Remark 1.13 (1) It is easy to verify that condition (2) in Theorem 1.11 implies condition (2) in Theorem 1.12. Moreover, using these two results it is easy to construct functions f such that $f \in SU(S_{n,k}) \setminus \text{int } SU(S_{n,k})$. For example, let $n=k=1$. If a function $f \in C[a,b]$ has 5 alternating extreme points, 3 of them in $[x_o,x_1)$ and 2 of them in $(x_1,x_2]$, then $f \in SU(S_{n,k}) \setminus \text{int } SU(S_{n,k})$ (where zero is the strongly unique best approximation of f).

(2) Theorem 1.11 remains true for the case when the functions f and the knots $x_1,\ldots,x_k$ of $S_{n,k}$ are variable, i.e. condition (2) in Theorem 1.11 characterizes "stability" of strong unicity also for this case. This follows from the observation in Remark 1.9 concerning variable (B,b) and from the fact that the basis functions of $S_{n,k}$ vary continuously with the knots (see Brosowski, Deutsch & Nürnberger [16]).

2. Unicity and strong unicity

In this section we investigate the relationship between unique and strongly unique solutions. It is well-known that they are not the same in general.

Example 2.1 Let $G = \text{span}\{g_1\}$, where $g_1(t) = t$, be a one-dimensional subspace of $C[-1,1]$ and $f \in C[-1,1]$ be defined by $f(t) = t+1$, if $t \in [-1,0]$, and $f(t) = 1-t$, if $t \in [0,1]$. Obviously zero is a best approximation of f, but not unique.

However, if we modify f in any small neighborhood of t=0 such that for the resulting function $\tilde{f}$ we have $\tilde{f}'(0) = 0$, then zero is a unique best approximation of $\tilde{f}$, but not strongly unique. This also shows that, since $\|\tilde{f}\|_\infty \approx \|\tilde{f}-g_1\|_\infty$, in practice we would not speak of a unique best approximation. Therefore, from the numerical point of view "strong unicity" may be considered as "unicity in practice".

We first give a necessary condition for unique solutions which should be compared with Theorem 1.2. (Here we use the convention that the set $\{t_o\}$ is a neighborhood of the point t_o.)

Theorem 2.2 If the optimization problem LM(σ) has a unique solution $x \in \mathbb{R}^N$, then for each $y \in \mathbb{R}^N$, $y \neq 0$, there exists a point $t \in N_{\sigma,x} \cup \{t_o\}$ such that for each neighborhood V of t there exists a point $v \in V$ with $\langle B(v),y\rangle < 0$.

Proof. Assume that there exists a vector $y \in \mathbb{R}^N$, $y \neq 0$, and a neighborhood W of $N_{\sigma,x} \cup \{t_o\}$ such that for all $t \in W$, $\langle B(t),y\rangle \geqq 0$. Then we set for all $\varepsilon > 0$, $x_\varepsilon = x - \varepsilon y$ and show that for sufficiently small $\varepsilon > 0$ the vector x_ε is a further solution of LM(σ), i.e. $\sigma \notin U$.
Since $\langle p,y\rangle = \langle B(t_o),y\rangle \geqq 0$, it follows that
$\langle p,x_\varepsilon\rangle = \langle p,x\rangle - \varepsilon\langle p,y\rangle \leqq \langle p,x\rangle$.
Furthermore, for all $t \in W$,
$\langle B(t),x_\varepsilon\rangle = \langle B(t),x\rangle - \varepsilon\langle B(t),y\rangle \leqq \langle B(t),x\rangle \leqq b(t)$.
Since $N_{\sigma,x}$ is contained in W, it follows that for all $t \in T\setminus W$, $\langle B(t),x\rangle - b(t) < 0$. Since the function $t \to \langle B(t),x\rangle - b(t)$ is continuous and $T\setminus W$ is compact, there exists a constant $K>0$ such that for all $t \in T\setminus W$, $\langle B(t),x\rangle - b(t) < -K$.
We can choose $\varepsilon > 0$ small enough such that for all $t \in T$, $\varepsilon|\langle B(t),x\rangle| \leqq K$. Then for all $t \in T\setminus W$,
$\langle B(t),x_\varepsilon\rangle = \langle B(t),x\rangle - \varepsilon\langle B(t),y\rangle \leqq b(t) - K + K = b(t)$.
This shows that x_ε is a further solution and proves Theorem 2.2.

The next result is an immediate consequence of Theorem 1.2 and Theorem 2.2, since, if T is finite, then the set $\{t\}$ is a neighborhood of $t\in T$.

Corollary 2.3 In finite optimization (i.e. T finite) unique and strongly unique solutions are the same, i.e. SU = U.

As a consequence of the next theorem we obtain a density result for arbitrary T.

Theorem 2.4 Let $\sigma=(p,B,b)\in Q$ and $x\in \mathbb{R}^N$ be a unique solution of $LM(\sigma)$. Then there exists a sequence (b_n) in $C(T)$ such that $\sigma_n=(p,B,b_n)\in Q$, $\sigma_n\to\sigma$ and x is a strongly unique solution of $LM(\sigma_n)$ for all n.

Proof. Let $x\in \mathbb{R}^N$ be a unique solution of $LM(\sigma)$. We set for all $t\in T$, $h(t) = \langle B(t),x\rangle - b(t)$.
Then $h(t) \leq 0$ for all $t\in T$ and $h(t) = 0$ for all $t\in N_{\sigma,x}$.
Let an integer n be given. We set $V_n = \bigcup\{B(t,1/n): t\in N_{\sigma,x}\}$, where $B(t,1/n)$ denotes the open ball with center t and radius $1/n$. By Urysohn's lemma there exists a function $z_n\in C(T)$ such that

$$z_n(t) = 0 \text{ for all } t\in\overline{V}_n,$$
$$z_n(t) = 1 \text{ for all } t\in T\setminus V_{n-1},$$
$$0 < z_n(t) < 1 \text{ for all } t\in V_{n-1}\setminus\overline{V}_n.$$

We set for all n, $h_n(t) = z_n(t)h(t)$, $b_n(t) = \langle B(t),x\rangle - h_n(t)$ for all $t\in T$ and $\sigma_n = (p,B,b_n)$.
Then $\|b-b_n\|_\infty \to 0$ and therefore $\|\sigma-\sigma_n\| \to 0$.
Let an integer n be given. Then for all $t\in T$,

$$\langle B(t),x\rangle - b_n(t) = h_n(t) \leq 0,$$

i.e. $x\in Z_{\sigma_n}$. Moreover, for all $t\in\overline{V}_n$,

$$\langle B(t),x\rangle - b_n(t) = h_n(t) = 0,$$

i.e. $\overline{V}_n \subset N_{\sigma_n,x}$. Now, let a vector $y\in \mathbb{R}^N$, $y\neq 0$, be given.
Since x is a unique solution of $LM(\sigma)$, it follows from Theorem 2.2 that $\langle B(t_o),y\rangle < 0$ or that there exists a point $t\in N_{\sigma,x}$

such that for each neighborhood V of t there exists a point $v \in V$ with $\langle B(v),y\rangle < 0$.
In the second case we choose V small enough such that $V \subset \overline{V}_n$. Since $\sigma \in Q$, we may assume that $\sigma_n \in Q$. Then it follows from Theorem 1.1 and Theorem 1.2 that x is a strongly unique solution of $LM(\sigma_n)$. This proves Theorem 2.4.

For fixed p and B we set $U_{p,B} = \{b\colon (p,B,b) \in U\}$.

<u>Corollary 2.5</u> Let a fixed vector $p \in \mathbb{R}^N$ and a fixed mapping $B \in C(T, \mathbb{R}^N)$ be given. Then $SU_{p,B}$ is a dense F_σ-set in $U_{p,B}$.

<u>Proof</u>. It follows from Theorem 2.4 that $SU_{p,B}$ is a dense subset of $U_{p,B}$. We show that $SU_{p,B}$ is the union of countably many closed sets in $U_{p,B}$.
Let $b \in C(T)$ be given such that $\sigma=(p,B,b) \in SU$ and let x be the strongly unique solution of $LM(\sigma)$. Moreover, let $K_b>0$ be the maximum of all numbers $K>0$ such that for all $y \in Z_\sigma$,

$$\langle p,y\rangle \geqq \langle p,x\rangle + K\,\|x-y\|_2 .$$

We set for all m,

$$F_m = \{b \in C(T)\colon (p,B,b) \in SU \text{ and } K_b \geqq 1/m\}.$$

Then obviously $SU_{p,B}$ is the union of the sets F_m.
Let an integer m be given. We show that the set F_m is closed in $U_{p,B}$. To do this, let (b_n) be a sequence in F_m converging to a function $b \in U_{p,B}$. We set $\sigma=(p,B,b)$ and $\sigma_n=(p,B,b_n)$ for all n. Let x (resp. x_n) be the unique solution of $LM(\sigma)$ (resp. $LM(\sigma_n)$). Then for all $y_n \in Z_{\sigma_n}$,

$$(*) \qquad \langle p,y_n\rangle \geqq \langle p,x_n\rangle + \frac{1}{m}\,\|x_n-y_n\|_2 .$$

It follows from Theorem 14 on p. 148 in Brosowski [4] that $x_n \to x$. Now, let $y \in Z_\sigma$ be given. It follows from Theorem 11,(2) on p. 142 in Brosowski [4] that there exists a sequence (y_n) such that $y_n \in Z_{\sigma_n}$ and $y_n \to y$. Taking limits it follows from (*) that

$$\langle p,y\rangle \geqq \langle p,x\rangle + \frac{1}{m}\,\|x-y\|_2 .$$

This proves Corollary 2.5.

Results of the above type have been given in Nürnberger & Singer [13] for best approximation.

Remark 2.6 (1) Examples in finite optimization and best approximation can be easily constructed which show that in general $SU_{p,B}$ (not even $U_{p,B}$) is not a dense subset of $L_{p,B} = \{b: (p,B,b) \in L\}$. For best approximation see Garkavi [5].
(2) In a further paper we will show that if p and B are fixed, then for every b with $\sigma=(p,B,b) \in L$ the optimization problem $LM(\sigma)$ has a unique (resp. strongly unique) solution if and only if there do not exist points $t_1,\ldots,t_{N-1} \in T$ and real numbers $a_1,\ldots,a_{N-1} \geq 0$ such that

$$-p = \sum_{i=1}^{N-1} a_i B(t_i).$$

Applications to various minimization problems will be given.

References

1. Bartelt, M.W. and McLaughlin, H.W. (1973) Characterizations of strong unicity in approximation theory. J. Approximation Theory 9, 255-266.
2. Blum, E. and Oettli, W. (1975) Mathematische Optimierung, (Springer, Berlin).
3. Brosowski, B. (1981) A refinement of the Kolmogorov-Criterion. In Constructive Function Theory (Sofia), 241-247.
4. Brosowski, B. (1982) Parametric Semi-Infinite Optimization, (Peter Lang, Frankfurt).
5. Garkavi, A.L. (1970) Almost Chebyshev systems of continuous functions. Amer. Math. Soc. Trans. 96, 177-187.
6. Glashoff, K. and Gustafson, S. (1978) Einführung in die lineare Optimierung, (Wissenschaftliche Buchgesellschaft, Darmstadt).
7. Hettich, R. and Zencke, H. (1982) Numerische Methoden der Approximation und semi-infiniten Optimierung, (Teubner, Stuttgart).
8. Krabs, W. (1975) Optimierung und Approximation, (Teubner, Stuttgart).
9. Meinardus, G. (1967) Approximation of Functions: Theory and Numerical Methods, (Springer, Berlin).

10. Nürnberger, G. (1980) Strong uniqueness of best approximations and weak Chebyshev systems. In Quantitative Approximation, R.DeVore and K. Scherer ed., (Academic Press), 255-266.

11. Nürnberger, G. (1982) A local version of Haar's theorem in approximation theory. Numer. Funct. Anal. and Optimiz. 5, 21-46.

12. Nürnberger, G., Strong unicity of best approximations: a numerical aspect. Numer. Funct. Anal. and Optimiz., to appear.

13. Nürnberger, G. and Singer, I. (1982) Uniqueness and strong uniqueness of best approximations by spline subspaces and other subspaces. J. Math. Anal. Appl. 90, 171-184.

14. Nürnberger, G., Schumaker, L.L., Sommer, M. and Strauß, H., Approximation by generalized splines, preprint.

15. Wulbert, D.E. (1971) Uniqueness and differential characterizations of approximations from manifolds. Amer. J. Math. 18, 350-366.

16. Brosowski, B., Deutsch, F. and Nürnberger, G. (1980) Parametric approximation. J. Approximation Theory 29, 261-277.

Günther Nürnberger, Fakultät für Mathematik und Informatik, Universität Mannheim, A 5, 6800 Mannheim 1, Federal Republic of Germany.

International Series of
Numerical Mathematics, Vol. 72

CONTINUOUS SELECTIONS IN CHEBYSHEV APPROXIMATION

Günther Nürnberger

Fakultät für Mathematik und Informatik, Universität Mannheim, D-6800 Mannheim, Federal Republic of Germany

Manfred Sommer

Mathematisch-Geographische Fakultät, Katholische Universität Eichstätt, D-8078 Eichstätt, Federal Republic of Germany

A survey on continuous selections for the set-valued metric projection onto finite-dimensional subspaces of C[a,b] is given. Some applications are discussed.

0. Introduction

Best approximation by finite-dimensional subspaces G of C[a,b] in the uniform norm $||\cdot||$ is considered. The set-valued mapping P_G which associates to every function f in C[a,b] the set of <u>best approximations</u>

$$P_G(f) = \{g \in G: ||f-g|| = d(f,G)\}$$

from G is called the <u>metric projection</u> onto G.

It is well-known that the metric projection $P_G: C[a,b] \to G$ is continuous (in the usual sense), if G is a Haar subspace (i.e. every $f \in C[a,b]$ has a unique best approximation from G). Prototypes of such spaces are subspaces of polynomials. However, there also exist important subspaces for which global uniqueness of best approximations is not given, e.g. subspaces of spline functions.

In these cases the metric projection is a set-valued mapping. There are various continuity concepts for such kind of mappings. In particular, we are interested in the following property. A continuous mapping $s: C[a,b] \to G$ is called <u>continuous selection</u> for P_G, if $s(f) \in P_G(f)$ for every $f \in C[a,b]$.

By Michael's selection theorem [25] lower semicontinuity is a sufficient condition for the existence of continuous selections. However, P_G is lower semicontinuous (if and) only if G is a Haar subspace (Blatter [4]). Thus Michael's theorem is not applicable to non-Haar subspaces. Although in these cases P_G is not lower semicontinuous, for certain subspaces G there actually exist continuous selections. In particular, if G is a spline subspace, then P_G admits a continuous selection if and only if the number of knots is less than or equal to the order of the spline functions ([31]).

The purpose of this article is to give a survey on the existence and uniqueness of continuous selections for P_G. Also pointwise-Lipschitz-continuous selections are considered.

Moreover, results on the relationship between the existence of continuous selections for the metric projection and other problems in approximation theory are presented: convergence of algorithms for computing best approximations, convergence of best L_p-approximations ($p \to \infty$), an extension of the well-known theorem of Mairhuber and existence of best approximations.

1. Continuous Selections

Let [a,b] be a compact real interval and let C[a,b] denote the space of continuous real-valued functions f on [a,b] endowed with the uniform norm $\|f\| = \sup\{|f(t)| : t \in [a,b]\}$. In the following G will always denote a finite-dimensional subspace of C[a,b].

The metric projection P_G is called <u>lower semicontinuous</u> (resp. <u>upper semicontinuous</u>), if for each $f \in C[a,b]$ and each open subset V of G with $P_G(f) \cap V \neq \emptyset$ (resp. $P_G(f) \subset V$) there exists a neighborhood U of f such that $P_G(\tilde{f}) \cap V \neq \emptyset$ (resp. $P_G(\tilde{f}) \subset V$) for every $\tilde{f} \in U$.

If dim G = n, then G is said to be a <u>Haar subspace</u>, if every nonzero function $g \in G$ has at most n-1 distinct zeros.

The well-known theorem of Haar states that every $f \in C[a,b]$ has a unique best approximation from G if and only if G is a Haar subspace (see e.g. Meinardus [24]).

A well-known continuity property of P_G is the following (see e.g. Singer [36]).

<u>Theorem 1.1</u>. The metric projection P_G is upper semicontinuous. In particular, if G is a Haar subspace, then P_G: $C[a,b] \to G$ is continuous (in the usual sense).

Michael's selection theorem [25] yields a sufficient condition for the existence of continuous selections for P_G.

<u>Theorem 1.2</u>. If P_G is lower semicontinuous, then there exists a continuous selection for P_G.

The following result shows that Michael's theorem is not applicable to the nontrivial case when G is a non-Haar subspace (Blatter [4], see also Blatter, Morris & Wulbert [5] and Brosowski & Wegmann [8]).

<u>Theorem 1.3</u>. The metric projection P_G is lower semicontinuous if and only if G is a Haar subspace.

(Note that the situation is completely different, if

[a,b] is replaced by a finite set (see Theorem 1.15).)

Despite of the negative statement in Theorem 1.3, for certain non-Haar subspaces G there actually exist continuous selections.

The first step was done by Lazar, Morris & Wulbert [22] who characterized those one-dimensional subspaces of C(T) (T an arbitrary compact Hausdorff space) for which there exist continuous selections.

Later a characterization of those arbitrary finite-dimensional subspaces of C[a,b] which admit continuous selections was given by the authors in a series of papers ([28], [30], [31], [37], [38], [40], [41], [42]).

Before stating this result we illustrate, how continuous selections are constructed. Let G be the subspace of linear spline functions in C[-1,1] with one fixed knot $x_1 = 0$.

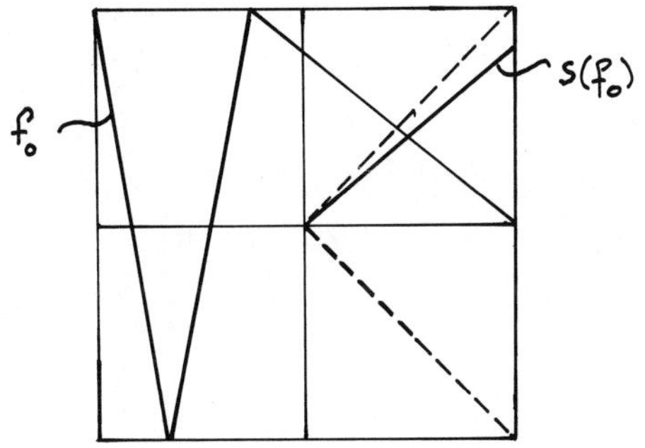

In this situation we have $P_G(f_0) = \{\alpha t_+ : -1 \leq \alpha \leq 1\}$, where the function $t_+ = \max\{0,t\}$.

Natural selections would be the extreme points of $P_G(f_0)$, namely $s(f_0) = \pm t_+$, or the "midpoint" $s(f_0) = 0$. If we choose for every $f \in C[a,b]$ such type of best approximations s(f), then the resulting selection s is <u>not</u> continuous. Also note that for all non-Haar subspaces G there does not exist a continuous selection s for P_G with the property that if $0 \in P_G(f)$, then $s(f) = 0$ for every $f \in C[a,b]$ ([27], see also Krüger [21]).

However, if we choose the uniquely determined best approximation $s(f_0)$ such that $\left(f_0 - s(f_0)\right)\big|_{[0,1]}$ has two extreme points with alternating sign and use a similar choice for all other functions $f \in C[-1,1]$, then the resulting selection s is actually continuous.

To state the above mentioned characterization we need the following notation.

If $\dim G = n$, then G is called a <u>weak Chebyshev subspace</u>, if every function $g \in G$ has at most $n-1$ sign changes (i.e. there do not exist points $a \leq t_1 < \ldots < t_{n+1} \leq b$ with $g(t_i)\, g(t_{i+1}) < 0$, $i = 1, \ldots, n$).

We denote by $Z(g) = \{t \in [a,b] : g(t) = 0\}$, by bd A the boundary of a subset A of $[a,b]$ and by $|A|$ the number of points in A.

The following property of weak Chebyshev subspaces was given in [41]:

If G is a weak Chebyshev subspace of $C[a,b]$, then there exists a certain set of knots $a = x_0 < x_1 < \ldots < x_s = b$ such that $G|_{[x_i, x_{i+1}]}$ is a Haar subspace or a weak Chebyshev subspace having the property that no nonzero $g \in G|_{[x_i, x_{i+1}]}$ vanishes on a subinterval of $[x_i, x_{i+1}]$, $i = 0, 1, \ldots, s-1$.

Now we are in position to state the following characterization due to the authors.

<u>Theorem 1.4</u>. The following statements (1) and (2) are equivalent:

(1) There exists a continuous selection s for P_G.

(2) (a) G is weak Chebyshev.

(b) No $g \in G$ has two separated zero intervals.

(c) For all $i, j \in \{0, 1, \ldots, s\}$, $i < j$, and all $g \in G$ with $g = 0$ on $[x_i, x_{i+j}]$,

$|\text{bd } Z(g)| \leq \dim \{g \in G : g = 0 \text{ on } [x_i, x_{i+j}]\}$.

(d) For all $i \in \{0, 1, \ldots, s-1\}$ and all nonzero $g \in G|_{[x_i, x_{i+1}]}$,

$|Z(g)| \leq \dim G|_{[x_i, x_{i+1}]}$.

As a consequence of Theorem 1.4 we obtain the following results for some special classes of subspaces of $C[a,b]$. We need the following notation.

A subspace G is called Z-subspace, if no nonzero function $g \in G$ vanishes on a nonempty open set. (Note that if G is a Z-subspace, then by a result of Garkavi [20],

$$\{f: f \text{ has a unique best approximation from } G\}$$

is a dense G_δ. Thus, if a continuous selection for P_G exists, it is unique.)

Corollary 1.5. If G is an n-dimensional Z-subspace, then the following statements are equivalent:

(1) There exists a continuous selection for P_G.

(2) G is weak Chebyshev and every nonzero function $g \in G$ has at most n distinct zeros.

In order to state the next result, spline functions must be defined.

Let $\Delta: a = x_0 < x_1 < \ldots < x_k < x_{k+1} = b$ be a partition of $[a,b]$. We denote by

$$S_m(\Delta) = \{s \in C^{m-1}[a,b]: s|_{[x_i, x_{i+1}]} \text{ is a polynomial of degree } m,\ i = 0,1,\ldots,k\}$$

the subspace of spline functions of degree m with k fixed knots.

Corollary 1.6. If $G = S_m(\Delta)$, then there exists a continuous selection s for P_G if and only if $k \leq m+1$.

Blatter & Schumaker [6], [7] studied the problem of uniqueness of continuous selections. Among others they obtained the following two results.

Theorem 1.7. If G is an n-dimensional Z-subspace, then there exists a unique continuous selection for P_G if and only if condition (2) in Corollary 1.5 is satisfied.

Theorem 1.8. If $G = S_m(\Delta)$, then there is never a unique continuous selection for P_G.

Blatt and the authors [2] showed that the selection constructed in Theorem 1.4 has even a stronger continuity property.

A selection s for P_G is called pointwise-Lipschitz-continuous, if for each $f \in C[a,b]$ there exists a constant $K(f)>0$ such that for all $\tilde{f} \in C[a,b]$,

$$\|s(f)-s(\tilde{f})\| \leq K(f) \cdot \|f-\tilde{f}\|.$$

Moreover, s is called quasi-linear, if for all $f \in C[a,b]$, $g \in G$ and scalars α, β,

$$s(\alpha f+\beta g) = \alpha\, s(f) + \beta g.$$

Theorem 1.9. The selection s from Theorem 1.4 is pointwise-Lipschitz-continuous and quasi-linear.

Freud [19] showed that every finite-dimensional Haar subspace G of C[a,b] has a pointwise-Lipschitz-continuous metric projection $P_G : C[a,b] \to G$.

Corollary 1.6 shows that for $G = S_m(\Delta)$ and $k > m + 1$ there does not exist a continuous selection for P_G. However, in this case the existence of a continuous selection was established in [44] provided that the approximation problem is considered on

$$I_\varepsilon = [a,b] \setminus \bigcup_{i=1}^{k} \{(x_i-\varepsilon, x_i) \cup (x_i, x_i+\varepsilon)\}$$

for sufficiently small $\varepsilon > 0$.

Theorem 1.10. If $G = S_m(\Delta)$, then there exists a pointwise-Lipschitz-continuous and quasi-linear selection s_ε for P_G: $C(I_\varepsilon) \to 2^G$.

Remark 1.11. Given $f \in C(I_\varepsilon)$, Strauß [47] constructed a particular best approximation $g_\varepsilon \in G = S_m(\Delta)$, the strict approximation in the sense of Rice [35]. The selection in Theorem 1.10 is defined by $s_\varepsilon(f) = g_\varepsilon$ for all $f \in C(I_\varepsilon)$.

Among others Brosowski et al. [9] studied the question how the set of best spline approximations to a fixed function depends on the knots which define the splines. They proved the following result.

Theorem 1.12. Let a fixed function $f \in C[a,b]$ be given. If $k \leq m+1$, then the set-valued mapping $\Delta \to P_{S_m(\Delta)}(f)$ admits a continuous selection.

Up to now we have considered the space C[a,b]. In the following we discuss a more general situation.

Theorem 1.4 shows that there is an intrinsic characterization of those n-dimensional subspaces G of C[a,b] which admit a continuous selection for P_G. However, if [a,b] is replaced by an arbitrary compact Hausdorff space T, a characterization is not known for $n > 1$. Lazar, Morris & Wulbert [22] gave a characterization for $n = 1$.

Theorem 1.13. If $G = \text{span}\{g_1\}$ is a one-dimensional subspace of C(T), then the following statements are equivalent:

(1) There exists a continuous selection for P_G.

(2) $|\text{bd}\, Z(g_1)| \leq 1$ and for each $t \in \text{bd}\, Z(g_1)$ there exists a neighborhood U of t such that g_1 has one sign on U.

In the particular case when G is an arbitrary finite-dimensional Z-subspace of C(T), there is a necessary condition for the existence of continuous selections. It was observed in [39] that the arguments in Brown [12] actually prove the following result.

Theorem 1.14. Let G be an n-dimensional Z-subspace of

$C(T)$. If there exists a continuous selection for P_G, then every nonzero $g \in G$ has at most n distinct zeros and at most $n-1$ zeros with sign changes.

(A function $f \in C(T)$ is said to have a zero $t \in T$ with a sign change, if for each neighborhood of t there exist points $t_1, t_2 \in U$ with $f(t_1)f(t_2) < 0$.)

It was shown in [29] and [39] that in particular cases the converse of Theorem 1.14 is true.

For $C(T)$, T a finite set, Brown [10] proved the following positive result (see also Blatter [4] for c_0).

Theorem 1.15. If T is a finite set and G is a subspace of $C(T)$, then P_G is lower semicontinuous and thus P_G admits a continuous selection.

Since lower semicontinuity is sufficient for the existence of continuous selections (see Theorem 1.2), but not necessary in general, Deutsch & Kenderov [15], [16] introduced various continuity concepts which are weaker than lower semicontinuity. In particular, they proved the following equivalence.

Theorem 1.16. If G is a one-dimensional subspace or a finite-dimensional Z-subspace of $C(T)$, then the following statements are equivalent:

(1) There exists a continuous selection for P_G.

(2) P_G is 2-lower semicontinuous.

(The metric projection P_G is called 2-lower semicontinuous, if for each $f \in C(T)$ and each $\varepsilon > 0$ there exists a neighborhood U of f such that for all $f_1, f_2 \in U$, $f_1 \neq f_2$,

$$\bigcap_{i=1}^{2} \{g \in G: d(g, P_G(f_i)) < \varepsilon\} \neq \emptyset.)$$

We conclude this section with a result on metric projections with respect to best L_p-approximation ($1 \leq p < \infty$).

The first statement is due to Lazar, Morris & Wulbert [22] and the second statement is well known (see e.g. Singer [36]).

Theorem 1.17. Let $1 \leq p < \infty$ and let G be a finite-dimensional subspace of $L_p[a,b]$.

(1) If $p = 1$, then there does not exist a continuous selection for P_G.

(2) If $1 < p < \infty$, then the metric projection $P_G : L_p[a,b] \to G$ is continuous (in the usual sense).

Further results on L_p-metric projections can be found in Deutsch [17].

2. Applications

There are certain relationships between the existence of continuous selections for the metric projection and other problems in approximation theory.

Brown [12] gave the following extension of the well-known theorem of Mairhuber [23].

Theorem 2.1. Let T be metrizable and let G be a finite-dimensional Z-subspace of C(T) with $\dim G > 1$ such that P_G admits a continuous selection. Then T is homeomorphic to a subset of the circle.

Mairhuber [23] proved Theorem 2.1 under the assumption that G is a Haar subspace.

Franchetti & Cheney [18] gave the following application of continuous selections to the existence of best approximations.

Theorem 2.2. For a finite-dimensional subspace G of C[a,b] the following statements are equivalent:

(1) For every function in $C(T, C[a,b])$ there exists a best approximation from C(T,G).

(2) There exists a continuous selection for P_G.

In this context compare Theorem 1.4 on the existence of continuous selections for P_G.

It was shown in [43], [44] that there is a relationship between the existence of continuous selections and the convergence of best L_p-approximations ($p\to\infty$).

For $f\in C[a,b]$ (resp. $f\in C(I_\varepsilon)$) and $1<p<\infty$ let $g_p(f)$ (resp. $g_{p,\varepsilon}(f)$) denote the unique best L_p-approximation of f from $G=S_m(\Delta)$ on $[a,b]$ (resp. on I_ε). (I_ε is the set defined on p. 7.)

<u>Theorem 2.3</u>. For $G=S_m(\Delta)$ the following statements hold:

(1) If $k\leqq m+1$ and $s:C[a,b]\to G$ is the continuous selection from Corollary 1.6, then $\lim_{p\to\infty} g_p(f)=s(f)$ for every $f\in C[a,b]$.

(2) If $s_\varepsilon:C(I_\varepsilon)\to G$ is the continuous selection from Theorem 1.10, then $\lim_{p\to\infty} g_{p,\varepsilon}(f)=s_\varepsilon(f)$ for every $f\in C[a,b]$.

<u>Remark 2.4</u>. Let T be a finite set and let G be a finite-dimensional subspace of C(T). For $f\in C(T)$, Rice [35] defines a particular best approximation from G, the so-called strict approximation, which can be considered as the "best among the best". Descloux [13] showed that the strict approximation is the limit of the sequence of best L_p-approximations of f ($p\to\infty$). Therefore by Theorem 2.3 the spline functions s(f) resp. $s_\varepsilon(f)$ can also be considered as the "best among the best" (see the figure on p. 4.)

There is also a relationship between the existence of continuous selections and the convergence of algorithms for computing best spline approximations developed by the authors [32] and by Strauß [48].

In [32] an algorithm of Remez type was developed: Let $n=\dim S_m(\Delta)$. Given a function $f\in C[a,b]$, in the p-th step a set M_p consisting of n+1 points $a\leqq t_{1,p}<\ldots<t_{n+1,p}\leqq b$ is deter-

mined for which there exists a unique spline function $s_p \in S_m(\Delta)$ and a unique real number λ_p such that

$$|(f-s_p)(t_{i,p})| = |\lambda_p|, \; i = 1, \ldots, n+1,$$

and $f-s_p$ alternates on M_p in a certain sense.

Strauß [48] developed an algorithm for computing the strict spline approximation: Given a function $f \in C[a,b]$, a knot-interval is determined and a best approximation of f on this interval is computed. In further steps best approximations are computed on larger knot-intervals.

For both algorithms the following convergence results were proved in [32] and [48].

If $k \leq m+1$, then the algorithms converge on $[a,b]$, and if $k > m+1$, then the algorithms converge on I_ε.

The relationship to the results in section 1 is, that precisely in these cases continuous selections for the metric projection exist (see Corollary 1.6 and Theorem 1.10).

Independently of [48] Stover [46] also defines strict approximations in the following sense. Let G be a finite-dimensional subspace of C[a,b] and let (T_q) be a sequence of increasing finite subsets of [a,b] which converges to a dense subset of [a,b]. For a given function $f \in C[a,b]$ let g_q be the strict approximation of f on T_q in the sense of Rice [35]. If there is a unique limit of the sequence (g_q), independent of the choice of (T_q), then this limit will be called the <u>strict approximation.</u>

In [46] the following result is proved.

<u>Theorem 2.5</u>. If G is a finite-dimensional subspace of C[a,b], then the following statements hold:
(1) Every function in C[a,b] has a strict approximation if and only if there exists a continuous selection for P_G.
(2) In particular, the strict approximations define a continuous selection.

3. References

1. Berdyshev, V.I. (1975) Metric projection onto finite-dimensional subspaces of C and L. Math. Zametki 18, 473-488 (Russian).

2. Blatt, H.-P., Nürnberger, G. and M. Sommer (1981-1982) A characterization of pointwise-Lipschitz-continuous selections for the metric projection. Numer. Funct. Anal. and Optimiz. 4 (2), 101-121.

3. Blatt, H.-P., Nürnberger, G. and M. Sommer (1980) Pointwise-Lipschitz-continuous selections for the metric projection, in "Approximation Theory III" (E.W. Cheney, ed.), Academic Press, New York, 223-228.

4. Blatter, J. (1967) Zur Stetigkeit von mengenwertigen metrischen Projektionen. Schr. Rheinisch-Westfälischen Inst. Instrum. Math. Univ. Bonn, Ser. A, Nr. 16.

5. Blatter, J., Morris, P.D. and D.E. Wulbert (1968) Continuity of the set-valued metric projection. Math. Annalen 178, 12-24.

6. Blatter, J. and L.L. Schumaker (1983) Continuous selections and maximal alternators for spline approximation. J. Approx. Theory 38, 71-80.

7. Blatter, J. and L.L. Schumaker (1982) The set of continuous selections of a metric projection in C(X). J. Approx. Theory 36, 141-155.

8. Brosowski, B. and R. Wegmann (1973) On the lower semicontinuity of the set-valued metric projection. J. Approx. Theory 8, 84-100.

9. Brosowski, B., Deutsch, F. and G. Nürnberger (1980) Parametric approximation. J. Approx. Theory 29, 261-277.

10. Brown, A.L. (1964) Best n-dimensional approximation to sets of functions. Proc. London Math. Soc. 14, 577-594.

11. Brown, A.L. (1971) On continuous selections for metric projections in spaces of continuous functions. J. Functional Anal. 8, 431-449.

12. Brown, A.L. (1982) An extension to Mairhuber's theorem. On metric projections and discontinuity of multivariate best uniform approximation. J. Approx. Theory 36, 156-172.

13. Descloux, J. (1963) Approximations in L^p and Chebyshev approximations. J. Soc. Indust. Appl. Math. 11, 1017-1026.

14. Deutsch, F. and J. Lambert (1980) On continuity of metric projections. J. Approx. Theory 29, 116-131.

15. Deutsch, F. and P. Kenderov (1980) When does the metric projection admit a continuous selection?, in "Approximation Theory III"(E.W. Cheney, ed.), Academic Press, New York, 327-333.

16. Deutsch, F. and P. Kenderov (1983) Continuous selections and approximate selections for set-valued mappings and applications to metric projections. SIAM J. Math. Anal. and Appl. 14.

17. Deutsch, F. (1983) A survey of metric selections, to appear in "Contemporary Mathematics" (ed. by R. Sine), Amer. Math. Soc. Providence.

18. Franchetti, C. and E.W. Cheney (1980) Best approximation problems for multivariate functions. Preprint.

19. Freud, G. (1958) Eine Ungleichung für Tschebyscheffsche Approximationspolynome. Acta. Sci. Math. (Szeged) 19, 162-164.

20. Garkavi, A.L. (1965) Almost Cebyshev systems of continuous functions. Izv. Vyss. Ucebn. Zaved. Matematika 45, 36-44 (Russian) [English transl. in: Amer. Math. Soc. Transl. 96 (1970), 177-187].

21. Krüger, H. (1980) A remark on the lower semicontinuity of the set-valued metric projection. J. Approx. Theory 28, 83-86.

22. Lazar, A.J., Morris, P.D. and D.E. Wulbert (1969) Continuous selections for metric projections. J. Functional Anal. 3, 193-216.

23. Mairhuber, J.C. (1956) On Haar's theorem concerning Chebyshev approximation problems having unique solutions. Proc. Amer. Math. Soc. 7, 609-615.

24. Meinardus, G. (1967) Approximation of Functions: Theory and Numerical Methods, Springer, New York.

25. Michael, E. (1956) Selected selection theorems. Amer. Math. Monthly 63, 233-237.

26. Nürnberger, G. (1975) Dualität von Schnitten für die metrische Projektion und von Fortsetzungen kompakter Operatoren. Dissertation, Erlangen.

27. Nürnberger, G. (1977) Schnitte für die metrische Projektion. J. Approx. Theory 20 (1977), 196-219.

28. Nürnberger, G. (1980) Nonexistence of continuous selections of the metric projection and weak Chebyshev systems. SIAM J. Math. Anal. 11, 460-467.

29. Nürnberger, G. (1980) Continuous selections for the metric projection and alternation. J. Approx. Theory 28, 212-226.

30. Nürnberger, G. and M. Sommer (1978) Weak Chebyshev subspaces and continuous selections for the metric projection. Trans. Amer. Math. Soc. 238, 129-138.

31. Nürnberger, G. and M. Sommer (1978) Characterization of continuous selections of the metric projection for spline functions. J. Approx. Theory 22, 320-330.

32. Nürnberger, G. and M. Sommer (1983) A Remez type algorithm for spline functions. Numer. Math. 41, 117-146.

33. Nürnberger, G., Schumaker, L.L., Sommer, M. and H. Strauß, Approximation by generalized splines. To appear.

34. Respess, J.R. and E.W. Cheney (1982) On Lipschitzian pro-

ximity maps, in "Nonlinear Analysis and Applications" (ed. by J.H. Burry and S.P. Singh), Vol. 80, Lecture Notes in Pure and Applied Math., Dekker, New York.

35. Rice, J.R. (1969) The Approximation of Functions, Vol. II, Addision-Wesly, Reading, Massachusetts.

36. Singer, I. (1974) The Theory of Best Approximation and Functional Analysis, CBMS 13, SIAM, Philadelphia.

37. Sommer, M. (1979) Continuous selections of the metric projection for 1-Chebyshev spaces. J. Approx. Theory 26, 46-53.

38. Sommer, M. (1980) Characterization of continuous selections for the metric projection for generalized splines. SIAM J. Math. Anal. 11, 23-40.

39. Sommer, M. (1982) Existence of pointwise-Lipschitz-continuous selections of the metric projection for a class of Z-spaces. J. Approx. Theory 34, 115-130.

40. Sommer, M. (1980) Nonexistence of continuous selections of the metric projection for a class of weak Chebyshev spaces. Trans. Amer. Math. Soc. 260, 403-409.

41. Sommer, M. (1980) Continuous selections for metric projections, in "Quantitative Approximation" (ed. by R. DeVore and K. Scherer), Academic Press, New York, 301-317.

42. Sommer, M. (1982) Characterization of continuous selections of the metric projection for a class of weak Chebyshev spaces. SIAM J. Math. Anal. 13, 280-294.

43. Sommer, M. (1983) L_p-approximations and Chebyshev approximations in subspaces of spline functions, in "Approximation and Optimization in Mathematical Physics" (ed. by B. Brosowski and E. Martensen), Peter Lang, Frankfurt, Bern, 105-139.

44. Sommer, M. Continuous selections and convergence of best L_p-approximations in subspaces of spline functions. To appear in Numer. Funct. Anal. and Optimiz.

45. Sommer, M. and H. Strauß (1977) Eigenschaften von schwach tschebyscheffschen Räumen. J. Approx. Theory 21, 257-268.

46. Stover, V. (1981) The strict approximation and continuous selections for the metric projection. Dissertation, San Diego.

47. Strauß, H. Characterization of strict approximations in subspaces of spline functions. To appear in Numer. Funct. Anal. and Optimiz.

48. Strauß, H. An Algorithm for the computation of strict approximations in subspaces of spline functions. To appear in J. Approx. Theory.

49. Wegmann, R. (1973) Some properties of the peak-set-mapping. J. Approx. Theory 8, 262-284.

Günther Nürnberger, Fakultät für Mathematik und Informatik, Universität Mannheim, 6800 Mannheim, Federal Republic of Germany

Manfred Sommer, Mathematisch-Geographische Fakultät, Katholische Universität Eichstätt, 8078 Eichstätt, Federal Republic of Germany